A Developer's Guide to Building Resilient Cloud Applications with Azure

Deploy applications on serverless and event-driven architecture using a cloud database

Hamida Rebai Trabelsi

BIRMINGHAM—MUMBAI

A Developer's Guide to Building Resilient Cloud Applications with Azure

Group Product Manager: Rahul Nair
Publishing Product Manager: Surbhi Suman
Senior Editor: Divya Vijayan/Romy Dias
Technical Editor: Arjun Varma
Copy Editor: Safis Editing
Project Coordinator: Ashwin Kharwa
Proofreader: Safis Editing
Indexer: Hemangini Bari
Production Designer: Aparna Bhagat
Marketing Coordinator: Nimisha Dua

First published: January 2023

Production reference: 1250123

Published by Packt Publishing Ltd.
Livery Place
35 Livery Street
Birmingham
B3 2PB, UK.

ISBN 978-1-80461-171-5

https://www.packtpub.com

To my mother, Rafika Boukari, and to the memory of my father, Mongi Rebai, for their sacrifices, their blessings, and the constant support that they have given to me. To my husband, Mohamed Trabelsi, for supporting me during my writing journey. To my son, Rayen Trabelsi, and my baby daughter, Eya Trabelsi, for their smiles every day, which always keep me motivated.

Foreword

On March 11, 2020, the World Health Organization officially declared a pandemic amid a global COVID-19 outbreak. As the world went into lockdown, it was imperative that businesses pivot and leverage innovative ways to continue their operations. We witnessed organizations employing creative techniques across a variety of industries. Many retail businesses provided online ordering and curbside pick-up when they were required to close their doors or limit the number of customers in-store.

When grocery store shelves were barren and struggling to restock, restaurants also leveraged their direct suppliers to offer grocery kits including produce, dairy, and raw meats to their customers. Fitness studios offered online virtual classes to keep their members active and engaged. Doctors provided online virtual appointments for their patients. Offices that traditionally required employees to be on-site transitioned to enable remote work using collaboration tools such as Microsoft Teams. What did they all have in common? They leveraged cloud services.

In his first earnings call during the pandemic, Satya Nadella, CEO of Microsoft, stated, "*We've seen two years of digital transformation in two months*." Enabling remote work, accelerating the digital transformation to the cloud, and building innovative solutions allowed organizations to not only survive but also thrive.

Three years later, the world has slowly gone back to a new normal – a hybrid world partaking in experiences online as well as in person. Yet, organizations are not slowing down. They are accelerating their digital transformation and application modernization initiatives. Gartner predicts that cloud spending will continue to increase through to 2025, with enterprise organizations continuing to invest in their shift to the cloud. I see this firsthand in my day-to-day work in leading a team of cloud solution architects that helps organizations to modernize and innovate in Azure.

What does this mean for you, whether you're an architect, developer, student, or aspiring technologist? As the demand for cloud-native skills continues to grow, now is the time to learn these crucial development skills, or risk getting left behind. A simple search for "cloud native" jobs on LinkedIn yields thousands of results. This is an in-demand skill that serves as a career accelerator. You have taken the first step by buying Hamida's book, which is a wise investment on your part.

Hamida draws on years of experience in developing cloud-native solutions in Azure and has distilled her expertise for you here. This is a comprehensive resource that goes beyond the question "what is cloud-native application development?" This is not another book on containers and orchestration services. Hamida provides you with the "how-to." She delivers the roadmap to build an end-to-end cloud-native application, from development to deployment.

You will walk away ready to put into practice everything that you learn within these pages. You will gain a deeper understanding of which cloud service to use, why, and how it all comes together to create and deploy a scalable, reliable, highly available application.

Good luck on your journey!

Lori Lalonde

Senior cloud solution architect manager, Microsoft

Contributors

About the author

Hamida Rebai Trabelsi has been working in the computing domain for over 12 years. She started her professional career in Tunisia working for **multinational corporations** (**MNCs**) as a software developer, then served as a .NET consultant at CGI, Canada. She is currently a senior advisor and information and solution integration architect at Revenu Québec, Canada. She has been awarded as Most Valuable Professional in Developer Technologies and Microsoft DevHeros by Microsoft and holds several Azure certifications. Besides being a Microsoft Certified Trainer and a member of the .NET Foundation, Hamida is a blogger, an international speaker, and one of the finalists in the Women in IT Award in Canada in 2019.

I have been overwhelmed by the support that I have received from all of the team. They encouraged me during this period, and I feel proud that I was able to finish this book and share my experience as a developer to start with Azure. To the memory of my father, Mongi Rebai, I offer this book to you and I hope that you were with me, sharing in this new adventure that is added to my professional career. Thank you to my mum, Rafika Boukari, for your sacrifices and your support. Thank you to my husband, Mohamed Trabelsi, for believing in me. Every day, I wake up and see my son, Rayen Trabelsi, and my baby girl, Eya Trabelsi, and I feel more motivated, so thank you for this energy.

About the reviewers

Ryan Mangan is a cloud and an end user computing technologist focusing on application migration and modernization. He's an international speaker on a wide range of topics, an author, and has helped customers and technical communities over the past decade. Ryan is a chartered fellow of the British Computer Society, a Microsoft **Most Valuable Professional** (**MVP**), as well as a VMware vExpert, and **Very Important Parallels Professional** (**VIPP**).

Stefano Demiliani is a Microsoft MVP in business applications, a **Microsoft Certified Solution Developer** (**MCSD**), an Azure Certified Architect, and an expert in other Microsoft-related technologies. His main activity is architecting and developing enterprise solutions based on the entire stack of Microsoft technologies (mainly focused on ERP and the cloud). He has worked with Packt Publishing on many IT books related to Azure cloud applications and Dynamics 365 Business Central and is a speaker at conferences around Europe. You can reach him on Twitter (@demiliani) or LinkedIn.

Table of Contents

3

Developing Event-Based and Message-Based Solutions   41

Part 2: Connecting Your Application with Azure Databases

4

Creating and Deploying a Function App in Azure   79

5

Develop an Azure Service Fabric Distributed Application   107

9

Working with Azure Cosmos DB to Manage Database Services   185

10

Big Data Storage Overview   211

Part 3: Ensuring Continuous Integration and Continuous Container Deployment on Azure

11

Preface

Azure is a Microsoft public cloud computing provider. It provides a range of cloud services including compute, network, storage, and analytics. Users will continue their journey in building cloud-oriented applications using serverless and event-driven technologies in Azure. , they will integrate their application with relational or non-relational databases and use Database-as-a-Service in Azure and build a CI/CD pipeline with Docker containers on Azure.

To remain competitive in the market and deliver software at a faster rate and reduced cost, companies with stable, legacy systems and growing volumes of data are modernizing their applications and accelerating innovation. However, many businesses struggle to meet modernization demands. This book will help you to build secure and reliable cloud-based applications on Azure using examples, and show you how to connect them to databases in order to overcome the application modernization challenges. The book will walk you through the different services in Azure, namely, Azure API Management using the gateway pattern, event-driven architecture, Event Grid, Azure Event Hubs, Azure message queues, Function-as-a-Service using Azure Functions, and the database-oriented cloud. At every step along the way, you'll learn about creating, importing, and managing APIs and Service Fabric in Azure, and how to ensure continuous integration and deployment in Azure to fully automate the software delivery process (the build and release process).

Who this book is for

This book is for cloud developers, software architects, system administrators, DBAs, data engineers, developers, and computer science students looking to understand the new role of software architects or developers in the cloud world. Professionals looking to enhance their cloud and cloud-native programming concepts will also find this book useful.

What this book covers

Chapter 1, *Introduction to Serverless Architecture, Event-Driven Architecture, and Cloud Databases*, covers the definition of serverless architecture, and the definition, life cycle, types, and roles of APIs inside an application and their ecosystem. This chapter covers also event-driven architecture and the database-oriented cloud.

Chapter 2, *API Management – Import, Manage, and Publish Your First API*, covers the different features of Azure API Management and defines the API gateway pattern. This chapter covers also API security.

Chapter 3, Developing Event-Based and Message-Based Solutions, covers event-based and message-based solutions. This chapter describes Event Grid, Azure Event Hubs, and Service Bus Queue and Topic. We will use a .NET application to send and receive messages to Service Bus Queue or Topic, and we will publish and subscribe events to Event Grid events using .NET application .

Chapter 4, Creating and Deploying a Function App in Azure, covers the basic concepts of Azure Functions and their hosting plan options, and the development of Azure Functions. This chapter also covers Durable Functions.

Chapter 5, Develop an Azure Service Fabric Distributed Application, covers the essential concepts of Azure Service Fabric, the main benefits, and Service Fabric application deployment locally or remotely to the cloud.

Chapter 6, Introduction to Application Data, covers data classification, data concepts, and the different concepts of relational and non-relational data in Azure. This chapter also covers modern data warehouse analytics.

Chapter 7, Working with Azure SQL Database, covers the provisioning and deployment of Azure SQL Database and Azure SQL Managed Instance. This chapter discusses the deployment of a single SQL database, database elastic pool, and SQL Managed instance. In this chapter, we will connect an Azure SQL database to an ASP.NET application.

Chapter 8, Working with Azure Storage, covers the different storage options that are available in Azure Storage services. This chapter discusses Azure storage accounts, including Azure Table storage, Azure Blob storage, Azure Disk storage, and Azure Files.

Chapter 9, Working with Azure Cosmos DB to Manage Database Services, covers the design and implementation of cloud-native applications using a multi-model NoSQL database management system. This chapter discusses Microsoft Azure Cosmos DB.

Chapter 10, Big Data Storage Overview, covers the different solutions for big data storage. This chapter gives an overview of Azure Data Lake Storage, Azure Data Factory, Azure Data Factory, Azure Synapse Analytics, and Azure Analysis Services.

Chapter 11, Containers and Continuous Deployment on Azure, covers the **continuous integration/ continuous delivery** (**CI/CD**) of containers on Azure. This chapter discusses the setting up of continuous deployment to produce your container images and orchestration.

To get the most out of this book

Having a good background in C#, ASP.NET Core, and Visual Studio (any recent version), and a basic knowledge of cloud computing and databases will be helpful when using this book.

Most of the examples presenting the solution reference are presented using .NET 6; basic knowledge in C# and .NET technology is essential to understand the described topics.

Software/hardware covered in the book	Operating system requirements
.NET 6 and .NET 7	Windows, macOS, or Linux
Docker Desktop	Windows, macOS, or Linux
Visual Studio 2022 (Community Edition)	Windows or macOS
Visual Studio Code	Windows, macOS, or Linux

We assume that you can install Visual Studio Code as an IDE. Visual Studio 2022 (the Community Edition) is enough to run the examples if you don't have access to the commercial license.

If you are using the digital version of this book, we advise you to type the code yourself or access the code from the book's GitHub repository (a link is available in the next section). Doing so will help you avoid any potential errors related to the copying and pasting of code.

Examples are used to explain the use of every Azure service.

Download the example code files

You can download the example code files for this book from GitHub at `https://github.com/PacktPublishing/A-Developer-s-Guide-to-Building-Resilient-Cloud-Applications-with-Azure`. If there's an update to the code, it will be updated in the GitHub repository.

We also have other code bundles from our rich catalog of books and videos available at `https://github.com/PacktPublishing/`. Check them out!

Download the color images

We also provide a PDF file that has color images of the screenshots and diagrams used in this book. You can download it here: `https://packt.link/LyxAd`.

Conventions used

There are a number of text conventions used throughout this book.

`Code in text`: Indicates code words in text, database table names, folder names, filenames, file extensions, pathnames, dummy URLs, user input, and Twitter handles. Here is an example: "The `az apim create` command is used to create the instance."

A block of code is set as follows:

```
var event1= new EventGridEvent(
   subject: $"New Patient: Hamida Rebai",
    eventType: "Patients.Registration.New",
    dataVersion: "1.0",
    data: new
   {
     FullName = "Hamida Rebai",
     Address = "Quebec, G2C0L6"
     }
);
```

When we wish to draw your attention to a particular part of a code block, the relevant lines or items are set in bold:

```
dotnet build
```

Any command-line input or output is written as follows:

```
Az group create -name packrg -location eastus
```

Bold: Indicates a new term, an important word, or words that you see onscreen. For instance, words in menus or dialog boxes appear in **bold**. Here is an example: "Select **Shared access policies** in the **Settings** section of the left-hand menu."

> Tips or important notes
> Appear like this.

Get in touch

Feedback from our readers is always welcome.

General feedback: If you have questions about any aspect of this book, email us at customercare@packtpub.com and mention the book title in the subject of your message.

Errata: Although we have taken every care to ensure the accuracy of our content, mistakes do happen. If you have found a mistake in this book, we would be grateful if you would report this to us. Please visit www.packtpub.com/support/errata and fill in the form.

Piracy: If you come across any illegal copies of our works in any form on the internet, we would be grateful if you would provide us with the location address or website name. Please contact us at copyright@packt.com with a link to the material.

If you are interested in becoming an author: If there is a topic that you have expertise in and you are interested in either writing or contributing to a book, please visit authors.packtpub.com.

Share your thoughts

Once you've read *A Developer's Guide to Building Resilient Cloud Applications with Azure*, we'd love to hear your thoughts! Scan the QR code below to go straight to the Amazon review page for this book and share your feedback.

https://packt.link/r/1804611719

Your review is important to us and the tech community and will help us make sure we're delivering excellent quality content.

Download a free PDF copy of this book

Thanks for purchasing this book!

Do you like to read on the go but are unable to carry your print books everywhere? Is your eBook purchase not compatible with the device of your choice?

Don't worry, now with every Packt book you get a DRM-free PDF version of that book at no cost.

Read anywhere, any place, on any device. Search, copy, and paste code from your favorite technical books directly into your application.

The perks don't stop there, you can get exclusive access to discounts, newsletters, and great free content in your inbox daily

Follow these simple steps to get the benefits:

1. Scan the QR code or visit the link below

https://packt.link/free-ebook/9781804611715

2. Submit your proof of purchase
3. That's it! We'll send your free PDF and other benefits to your email directly

Part 1: Building Cloud-Oriented Apps Using Patterns and Technologies

In this part of the book, we will introduce the serverless architecture, APIs, event-driven architecture, and the database-oriented cloud.

This part comprises the following chapters:

- *Chapter 1, Introduction to Serverless Architecture, Event-Driven Architecture, and Cloud Databases*
- *Chapter 2, API Management – Import, Manage, and Publish Your First API*
- *Chapter 3, Developing Event-Based and Message-Based Solutions*

1

Introduction to Serverless Architecture, Event-Driven Architecture, and Cloud Databases

This first chapter introduces the book's content, which includes the modernization of apps using **application programming interfaces (APIs)**, **event-driven architecture**, functions, and **Service Fabric**, and connecting them with a database.

Modernization is not just reserved for companies that have been operating stable systems for years and whose data volume has evolved. Organizations want to remain profitable and not be left behind by not modernizing technology and taking advantage of changes in the market due to world-changing factors, such as Covid-19.

Modernization is about accelerating innovation and reducing costs. Modernization is not just about the app but also the data.

In this chapter, we're going to cover the following main topics:

- Understanding serverless architecture
- Understanding the role of APIs inside an application and their ecosystem
- Understanding event-driven architecture
- Exploring cloud databases

Understanding serverless architecture

Serverless architecture is a software design pattern where applications are hosted by a third-party service, allowing developers to build and run services without having to manage the underlying infrastructure.

Applications are divided into separate functions that can be called and scaled individually.

Developers implement and deploy application-only code and can run applications, databases, and storage systems hosted in servers provisioned by cloud providers.

We will understand the role of APIs inside an application and the ecosystem. To start modernizing legacy applications without any modification, we can use APIs to interact with existing code. These APIs will run in the cloud, but if we need to add more features to the application to modernize it, we will encapsulate the data and the functions in services via an API. This encapsulation phase is a technique that consists of reusing legacy software components. The goal is to keep the code in its environment and connect it to the new presentations via encapsulation to access the different layers through an API. APIs permit you to define common protocols (rules) to expose data from an application with the possibility of decoupling the presentation layer.

Technically, the encapsulation works with wrapper technology, providing a new interface for a legacy component. This component will be easily accessible from the rest of the software components.

These minimal modifications reduce the risk of destabilizing the application. Even if encapsulation is a fast, effective, and less expensive solution, it will not solve the present problems related to the difficulties of maintenance or upgrading.

Because we are talking about the role of APIs in digital transformation and their life cycle, we need to define them.

API definition

APIs form a bridge between different applications to ensure communication and the exchange of information between them. They can even define the behavior of these applications.

APIs are considered connectors because they provide the ability for disparate applications to exchange information.

The information exchanged is generally data; for example, if we make a request in a mobile application, the data will be sent to a server that performs the reading and then sends back a response in a format readable in JSON.

Sometimes, microservices are compared with APIs. We often talk about the relationship between APIs and microservices, but there is also a difference between them. Microservices are a style of architecture that divides an application into a set of services; each service presents a very particular domain, for example, an authentication service or another service for the management of the products. On the other hand, an API is a framework (a structure that you can build software on) used by developers to provide interaction with a web application. Alternatively, microservices can use an API to provide communication between services. But how does an API work?

Applications that send requests are called **clients,** and applications that send responses are called **servers**. Using bees as an example, the hive is the client, the flower is the server, and the bee is the communication path (REST API requests).

The API life cycle

The API manager – usually the enterprise architect or API product manager – manages the API life cycle.

The API life cycle consists of the following three main phases:

- **The creation phase**: This consists of creating and documenting the API. There are some key aspects of API creation to consider; the API will use or reuse backend resources (service or business capability implementation). These resources are typically available as RESTful services, **Simple Object Access Protocol (SOAP)**-based web services, or even **Advanced Message Queuing Protocol (AMQP)**-compliant message brokers.

- **The control phase**: This consists of applying the security policies necessary to ensure that the exchanges are secure.

- **The consumption phase**: This consists of publishing the API so that we can consume and monetize it.

After understanding the API life cycle and the different phases, we will next look at the important role of an API in terms of communication between applications.

An APIs role

A set of rules ensures the communication of APIs by defining how applications or machines can communicate, so the API is an intermediate bridge between two applications wanting to communicate with each other.

API types

A web API is an API that can be accessed via the HTTP protocol. There are many API protocols/specifications, as follows:

- Open APIs
- Partner APIs
- Internal APIs
- Composite APIs

- **Representational State Transfer (REST)**

- SOAP

- **Extensible Markup Language Remote Procedure Call (XML-RPC)**

- **JavaScript Object Notation Remote Procedure Call (JSON-RPC)**

The most popular type of API is **RESTful** because it has several advantages in terms of flexibility when creating an API that meets the client's needs and dependencies because the data is not bound to methods or resources. A RESTful API supports different data formats, such as `application/json`, `application/xml`, `application/x-wbe+xml`, `multipart/form-data`, and `application/x-www-form-urlencoded`.

RESTful APIs take advantage of existing protocols – for example, web APIs take advantage of the HTTP protocol.

So far, in this chapter, we have talked about encapsulation to access the different layers through an API as a technique used for legacy system modernization. We have presented the API life cycle and API roles, which encompasses several roles to ensure communication, and we have identified the API types, such as RESTful APIs, which are the most popular. In the next section, we will present the event-driven architecture used to improve agility in complex applications.

Understanding event-driven architecture

Event-driven architecture is a software architecture that uses events in order to be able to communicate between decoupled services. It is a pattern for designing applications that are loosely coupled and is used in modern applications built with microservices. When consumers are listening to an event, which could be a status change or an update, event producers are not able to know which event consumers are listening to and do not even know the consequences of its occurrence.

In an event-driven architecture, we have the following three key components:

- **Event producers**: These generate a stream of events

- **Event routers**: These manage event delivery between producers and consumers

- **Event consumers**: These listen to the events

The following diagram illustrates these components:

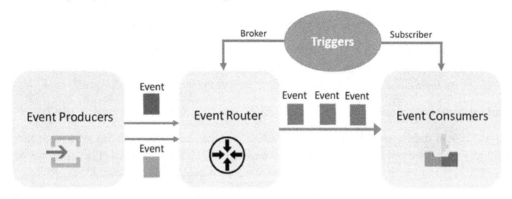

Figure 1.1 – Event-driven architecture

The source of an event is triggered by internal or external inputs. They can be generated by either a user (by clicking or using keyboard input, for example), an external source (such as a sensor output), or through a system (such as loading a program).

In an event-driven architecture, we can use event streaming or a publisher/subscriber model. But what is the difference between event streaming and a publisher/subscriber model?

- **Publisher/subscriber model**: This provides a framework that enables message exchanges between publishers and subscribers. This pattern involves the publisher and the subscriber and depends on a message broker that reroutes messages from the publisher to the subscriber.

- **Event streaming**: When a stream of events is published to a broker, the clients are able to subscribe to the stream and join at any time; they have access to them and can consume multiple preferred streams, and they are able to read from any part and advance their position. The events are always written in a log file.

Event-driven architectures are recommended to improve agility and move quickly. They are used in modern applications, mainly microservices, or in any application that includes several decoupled components. When adopting an event-driven architecture, you may need to rethink how you view your application design.

In this section, we have explored event-driven architecture and the different key components.

Exploring cloud databases

A **cloud database** is a database service created and accessed through a cloud platform.

A cloud database is a collection of information, structured or unstructured, that is hosted in a private, public, or hybrid cloud computing infrastructure platform. There is no structural or conceptual difference between a cloud or on-premises database, it's just the location that's different.

Cloud databases are divided into two broad categories: relational and non-relational.

As in the case of databases with traditional ancestors, we have the same definition for a **relational database**, which is written in **Structured Query Language** (**SQL**). It is composed of tables organized in rows and columns with relationships between them, called **fields**. This relationship is specified in a data schema.

Non-relational databases, also called **NoSQL**, use a different storage concept based on documents. They do not use a table model to store content as in the traditional approach; they use a single document instead.

A non-relational database is recommended for unstructured data – for example, for social media content, photos, or video storage. There are two models of cloud database environments: traditional (which we discussed earlier) and **database as a service** (**DBaaS**).

For the first case, we can host a virtual machine, install the cloud **database management system** (**DBMS**), and the database runs on this machine, so the management and monitoring of the database are managed by the organization. On the other hand, the DBaaS model is a paid subscription service in which the database runs on the physical infrastructure of the cloud service provider.

Azure offers a set of fully managed relational, NoSQL, and in-memory databases:

- **Azure SQL Database**: This is used for applications that scale with intelligent, managed SQL databases in the cloud

- **Azure SQL Managed Instance**: This is used to modernize your SQL Server applications with a managed, always up-to-date SQL instance in the cloud

- **SQL Server on Azure Virtual Machines**: This is used to migrate SQL workloads to Azure while maintaining full SQL Server compatibility and OS-level access

- **Azure Database for PostgreSQL**: This is used to build scalable, secure, fully managed enterprise-grade applications using open source PostgreSQL, scale PostgreSQL with single-node and high performance, or move your PostgreSQL and Oracle workloads to the cloud

- **Azure Database for MySQL**: This used to provide high availability and elastic scaling for your open source mobile and web apps with the managed community MySQL database service or move your MySQL workloads to the cloud

- **Azure Database for MariaDB**: This is used to build applications anywhere with guaranteed low latency and high availability at any scale, or move Cassandra, MongoDB, and other NoSQL workloads to the cloud

- **Azure Cache for Redis**: This is used to run fast and scalable applications with open source compatible in-memory data storage

- **Azure Database Migration Service**: This is used to accelerate your move to the cloud with a simple, self-paced migration process

- **Azure Managed Instance for Apache Cassandra**: This is used to modernize existing Cassandra data clusters and apps and enjoy flexibility and freedom with the Managed Instance service

Summary

If you are building a new application or are in the process of modernizing a legacy application, you need to understand these architectures: serverless architecture, APIs, and event-driven architecture. These architectural patterns are critically important to master.

This chapter was about serverless architecture, API definition, types, life cycles, communication between applications or machines, event-driven architecture, and cloud databases.

In the next chapter, you will learn how to deploy web APIs and the explore function of the API management service.

Further reading

If you need more information about serverless architecture, you can check out this link: `https://azure.microsoft.com/en-us/resources/cloud-computing-dictionary/what-is-serverless-computing/` and this e-book: `https://learn.microsoft.com/en-us/dotnet/architecture/serverless/`.

You can check out event-driven architecture documentation at these links: `https://learn.microsoft.com/en-us/azure/architecture/guide/architecture-styles/event-driven` and `https://learn.microsoft.com/en-us/azure/architecture/reference-architectures/serverless/event-processing`.

Questions

1. What are the three key components of event-driven architecture?

2. What are the different Azure database services?

API Management – Import, Manage, and Publish Your First API

In this chapter, we will explore the different features of **Azure API Management**. We will understand the **API Gateway pattern** and the management of **API calls**. We will also publish the web APIs of the healthcare solution to Azure API Management, securing the API using subscriptions and certificates and creating a backend API.

In this chapter, we're going to cover the following main topics:

- Technical requirements
- The API Gateway pattern
- Exploring the API Management service
- Securing the API
- Exercise 1 – creating a backend API and deploying APIs
- Exercise 2 – using Azure API Management to proxy a public API

Technical requirements

Before you start exploring our solution, you will need a code editor for **C#**. Microsoft offers code editors and **integrated development environments (IDEs)**, such as the following:

- Visual Studio Code for Windows, macOS, or Linux
- Visual Studio 2022 for Windows
- Visual Studio 2022 for Linux
- GitHub Codespaces

As a prerequisite, we need an Azure account where we are able to deploy our APIs on Azure App Service and Azure API Management.

The healthcare solution presented in this chapter uses the .NET 6 framework.

The GitHub repository for this book has solutions using full application projects related to the healthcare domain, and they will be used in the next chapters:

`https://github.com/didourebai/developerGuideForCloudApps`

The API Gateway pattern

When developing applications based on complex or large microservices established on multiple client applications, it is recommended to use the API Gateway pattern.

Definition

The API Gateway pattern is an integration pattern for clients that communicate with a system service, designed to provide a single abstraction layer between the underlying services and the customer's needs. This is the single entry point for all clients. It is similar to the Façade pattern of **object-oriented programming (OOP)** design, but in this case, it's part of a distributed system.

The API Gateway pattern is often referred to as *Backend for Frontend* because its implementation is based on the needs of the client application. The pattern provides a reverse proxy whose purpose is to redirect or route client requests to internal microservices endpoints.

When should you consider using the API Gateway pattern? The following list specifies when it is suitable:

- A synchronous response is issued by the system making the call
- Dependencies between microservices are manageable and do not change over time
- No latency requirement; there will be no critical issues if there are slow responses
- There is a need to expose an API for the purpose of performing data collection from various microservices

After defining the API Gateway pattern, we will discover, in the next section, the use case that we will use during this chapter.

Use case

Let's start by looking at a problem relating to the direct communication between the client app and microservices. The following figure shows our solution, which contains several frontend and backend services that are directly interacting with the client applications.

In our scenarios, the client apps must send requests directly to the microservices. How does the client know which endpoints to call? What happens if we introduce a new service, for example, or if we decide to make a change and refactor an existing service?

How do services handle authentication or **Secure Sockets Layer** (**SSL**) termination, data transformations, and dynamic request dispatching?

How can client apps communicate with services that use different protocols? There are some potential problems when we expose services directly to clients:

- The client must track multiple endpoints that have been implemented to ensure resilient fault handling, resulting in a complex client code implementation.

- Coupling between the client and backend. This makes it more difficult to maintain the client and also more difficult to refactor services.

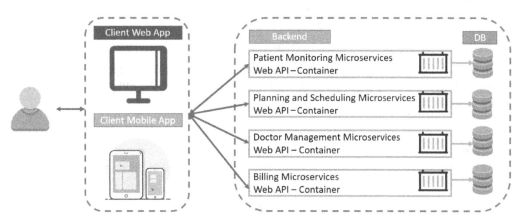

Figure 2.1 – Scenario using a direct client-to-microservice communication architecture

All these challenges can be resolved by implementing API Gateway pattern to decouple clients from services because having an intermediate level or tier of indirection (gateway) can be convenient for microservices-based applications. The API gateway sits between the client apps and the microservices.

The following figure shows a simplified architecture based on microservices and interaction with customers through the integration of a custom API gateway:

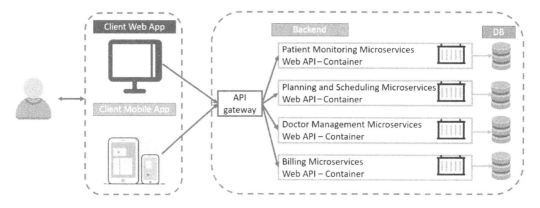

Figure 2.2 – Scenario using an API gateway implemented as a custom service

Along with the advantages of using the API Gateway pattern, there are a few disadvantages that should not be overlooked:

- It can be complex to set up because the API gateway is yet another additional part to be developed, deployed, and managed separately from the microservices and the complete solution.

- There is a possibility of latency during communications. The response time may be slower than a simplified architecture due to the additional network hop through the API gateway – however, for most applications, the additional roundtrip is not costly.

Organizations are adopting microservices because of the architecture's underlying scalability and flexibility. However, an API gateway is required to fully exploit the benefits of a microservices strategy. Azure API Management includes an API gateway, and we will explore it in the next section.

Exploring the API Management service

We will describe, in this section, the components of API Management and their functions, and we will explore the products that use APIs in API Management.

API Management components

Azure API Management is a hybrid, portable, multi-cloud management platform for APIs, providing as core functionality the assurance of a successful API program through developer engagement, business insights, analytics, security, and protection. Each API can include one or more operations and can be added to one or more products.

How can an API be used in API Management?

The use of an API by developers is really simple and interactive. They must subscribe to a product that includes this API, and then they can call the desired operation of the API by ensuring enforcement of usage policies that may be in effect.

APIs provide a simplified experience when integrating applications, enabling the reuse and accessibility of new product and service data on a universal scale.

With the increase of and, above all, the growing dependency on the publication and use of APIs, each company must manage these APIs as first-rate assets meeting several criteria.

These criteria are the API life cycle; the diversification and complexity of the backend architecture of API consumers; the protection, acceleration, and observation of the operation of APIs; as well as the consumption of APIs by the different users and the securing of these hosted services that are in or outside Azure. All these criteria are supported by Azure API Management.

Azure API Management is made up of the following components:

- API gateway
- Azure portal
- Developer portal

The following diagram presents the different components of Azure API Management:

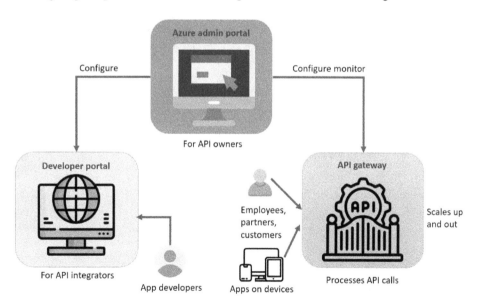

Figure 2.3 – Azure API Management components

We have explored the different ways to use API Management and its different components. We will look at the API gateway next.

API gateway

As defined before, the API gateway is the endpoint. It accepts API calls made by developers, and they will be routed to one or more backends depending on the API's internal design. A verification of **API keys**, **JSON Web Tokens** (**JWTs**), **certificates**, and other credentials are also performed at the endpoint level. If your API is called without limit, which slows and generates disturbances in the system, the verification of credentials can also be managed by applying the quotas of use and the limits of flow. We can perform API transformations on the fly without any code changes to the API gateway.

Metadata will be logged for analysis purposes, as well as backend response caching, where configured.

In the following screenshot, we can see the gateway URL displayed when we create an instance of the API Management service in Azure:

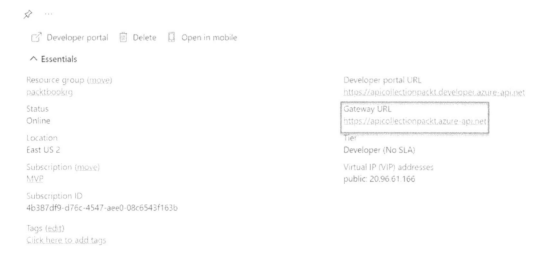

Figure 2.4 – Gateway URL in Azure API Management

We will start by creating an API Management service using the Azure portal.

The Azure portal

The Azure portal includes more than 200 cloud products and services, and API Management is an administrative interface where you can configure an API program.

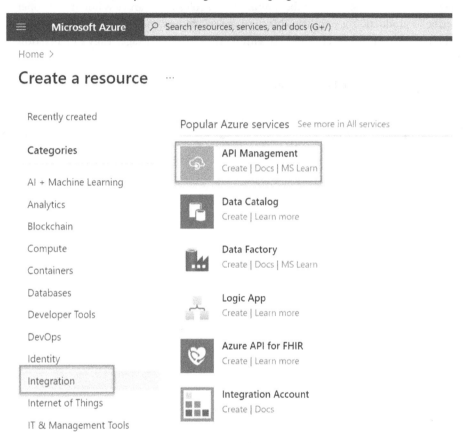

Figure 2.5 – API Management in the Azure portal

Through API Management, we can define or import an API schema, group APIs into products, manage users, get insights from analytics, or even configure access or consumption policies, such as quotas or the transformations performed on the APIs.

Developer portal

When we create a new API Management instance, an open source developer portal is automatically generated. It is used by developers and is a website that can be personalized, including the documentation of your APIs.

Developers can customize the look of the portal by adding custom content, for example, styling or even their own branding. We can extend the developer portal further through self-hosting.

Application developers use the open source developer portal to learn about their APIs, read the API documentation, or try out an API via the interactive console. They can view and invoke operations and subscribe to products. They are also able to create an account and subscribe to get API keys and manage them, they can access a dashboard with metrics on their usage. They can also download API definitions.

The developer portal URL can be found on the Azure portal dashboard for your API Management service instance.

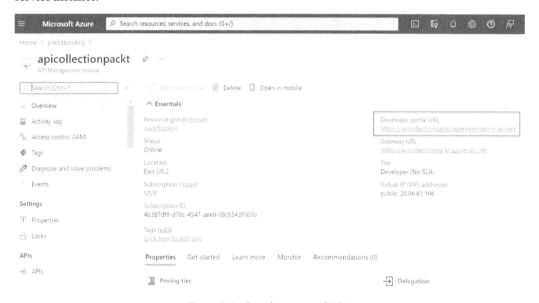

Figure 2.6 – Developer portal URL

Developers will need to have user accounts in an API Management service instance. The administrator registers developers or invites a developer to join to register from the developer portal, and each developer is a member who can belong to one or more groups; they are able to subscribe to products that are visible to these groups.

To manage user accounts in Azure API Management, follow these steps:

1. Open your Azure API Management instance and select **Users** under **Developer portal**.

2. To create a user, select **+ Add** and provide the information required (first name, last name, email address, ID, and password), as presented in the following screenshot:

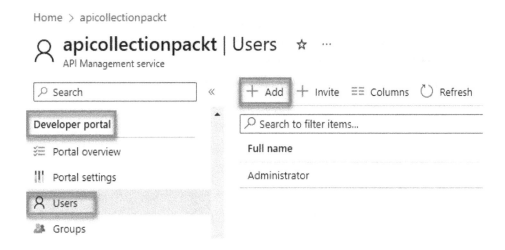

Figure 2.7 – Create a new developer in Azure API Management

By default, newly created developer accounts have an active status and are assigned to the Developers group. An active developer account can be used to access any API to which they have a subscription.

We can invite developers to the developer portal access by sending a notification to them from the developer portal. Follow these steps to send an invite:

1. In **Users** under **Developer portal**, select **Invite**.

2. Fill in all the information required to invite a user. An email will be sent to the developer. A customized template can be used for the email. If the user accepts the invitation, the account will be activated. Note that the invitation link is valid for 48 hours.

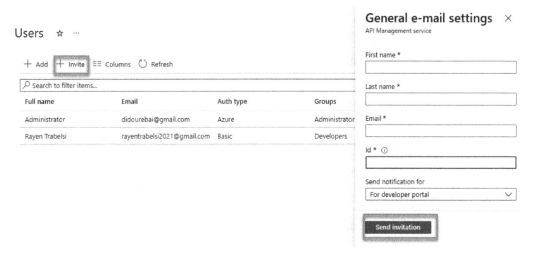

Figure 2.8 – Invite a developer

To disable a developer account, in order to block their access to the developer portal or call any APIs, select **Block** as presented in *Figure 2.9*. To reactivate a blocked developer account, select **Activate**. If we want to delete an existing active user, we select **Delete**.

These options are shown in the following screenshot:

Figure 2.9 – Block or delete a developer account in Azure API Management

A user is associated with a group. A group is used to provide product visibility management for each developer. API Management has the following immutable system groups:

- **Administrators**: Azure subscription administrators are members of this group. Administrators manage the API Management service instance. They create the APIs, operations, and products used by developers.

- **Developers**: This group is for authenticated developer portal users who build applications using APIs. They are allowed to access the developer portal and are able to build applications that call the operations of an API.

- **Guests**: Unauthenticated users of the developer portal, such as potential customers visiting the developer portal of an API Management instance, belong to this group. An administrator can grant them read-only access, such as the ability to view APIs without calling them.

Administrators can create custom groups or make use of external groups in Azure Active Directory tenants.

Products

Products are one or more APIs in API Management that are configured with a title, a description, and terms of use with access restrictions: open or protected. The only difference between open and protected is that protected products must be subscribed to before their use or consumption by developers; open ones can be used without a subscription. Configuring subscription approval is done at the product level itself, requiring admin approval or auto approval.

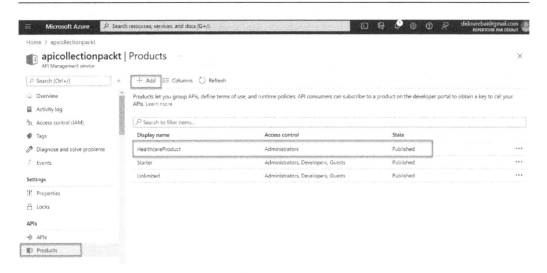

Figure 2.10 – Products in Azure API Management

One of the most powerful API Management features that allow the Azure portal to modify API behavior through configuration is **policies**.

Policies present a collection list of statements that can execute sequentially on the request or response of an API. The most well-known declarations include format conversion from XML to JSON.

Creating and publishing a product

In Azure API Management, a product includes one or more APIs, a usage quota, and terms of service. Once the product is released, developers can subscribe and start using these APIs.

To create and publish a product, you need to follow these steps:

1. Sign in to the Azure portal and navigate to your API Management instance.
2. In the left navigation pane, select **Products**; after that, select the + **Add** button.

3. In the **Add product** window, fill in the fields, as displayed in the following screenshot, to create your product:

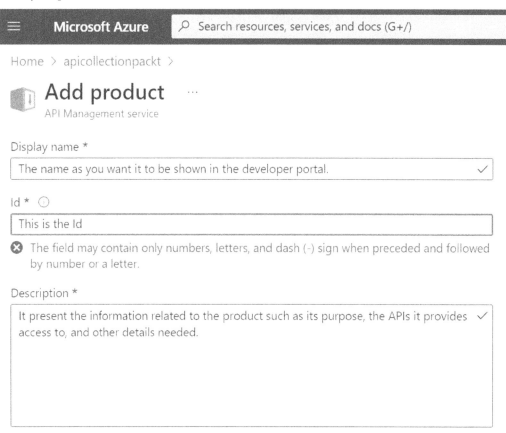

Figure 2.11 – Add product part 1

If we scroll down the **Add product** window, we will see what is shown in the following screenshot. We need to fill in more information and select the checkboxes to define whether the product can be published or not, and so on:

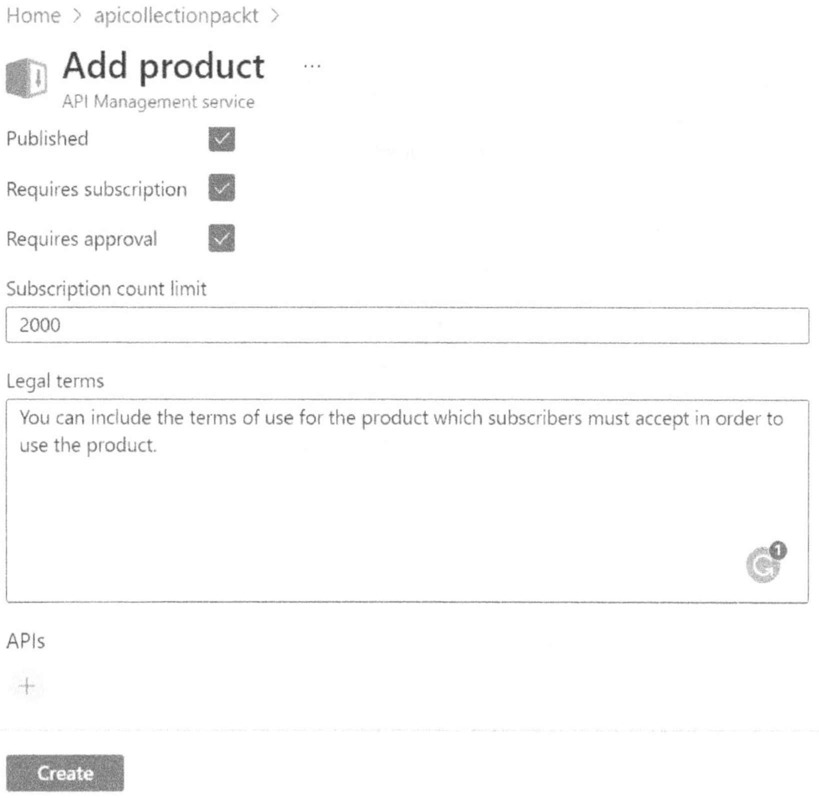

Home > apicollectionpackt >

🗃 **Add product** ⋯
API Management service

Published   ✓

Requires subscription   ✓

Requires approval   ✓

Subscription count limit

2000

Legal terms

You can include the terms of use for the product which subscribers must accept in order to use the product.

APIs

+

Create

Figure 2.12 – Add product part 2

If **Published** is checked, the product will be published, and the APIs in this product can be called; else, the unpublished product will be visible only to the Administrators group.

If **Requires subscription** is checked, the user is required to subscribe before using the product because the product is protected, and a subscription key must be used to access the product's APIs; else, the product is open and doesn't require a subscription key.

If **Requires approval** is checked, the administrator will review and accept or reject the subscription attempts to this product, else the subscription attempts will be approved automatically.

Subscription count limit applies a limit to the number of simultaneous subscriptions and is optional.

Under **APIs**, we can select one or more APIs to add them to the product to create. It is optional because you can also add APIs after creating the product.

4.   In the end, select **Create** to create your new product.

Once the new product is created, it can be used by developers or any application.

Securing the API

When publishing APIs via API Management, access to these APIs is secured by using subscription keys. Developers must include a valid subscription key in HTTP requests when calling an API; otherwise, these calls will be rejected by the API Management gateway. However, the transmission to the backend is not ensured.

If a developer wants to consume published APIs, a subscription is required. Developers who want to consume the published APIs must include a valid subscription key in HTTP requests when calling those APIs. But the calls can be rejected immediately by the API Management gateway or will not be forwarded to the backend services without a valid subscription key. They can get a subscription without approval from API publishers, although API publishers can even create subscriptions directly for API consumers.

Several API access security mechanisms are supported for Azure API Management Service, such as OAuth 2.0, client certificates, and IP allow lists.

Subscriptions and keys

A subscription key is a unique auto-generated key. This key can be passed through in the headers of the client request or as a query string parameter.

A subscription offers full and detailed control over authorizations and policies. A key is linked directly to a subscription and can be extended to different domains.

Each application must include the key in every request when calling a protected API.

Regeneration of subscription keys is easy and possible if we want to share a key with another unauthorized user.

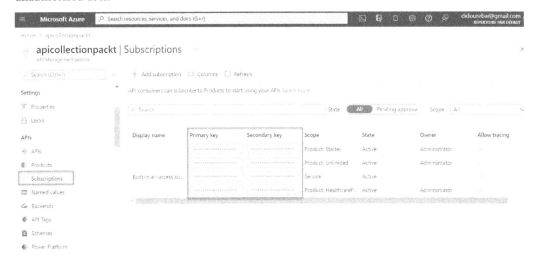

Figure 2.13 – Subscription in Azure API Management

The preceding screenshot shows that a subscription has two keys, a primary and a secondary key, which makes it easier to regenerate it. For example, in the case of modifying the primary key, in order to avoid downtime, you can use the secondary key in your applications.

Going back to the product that already includes our APIs where subscriptions are enabled, each customer must provide a key during the product API call.

Developers will submit a subscription request to obtain a key. If this request is approved, then the subscription key will be sent in a secure (encrypted) way.

The process of calling an API with the subscription key

The default header name is `Ocp-Apim-Subscription-Key`, and the default query string is `subscription-key`. Applications making calls to subscription-protected API endpoints must include a valid key in all of their HTTP requests passed in the request header or as a query string in the URL.

If we want to test the API calls, we can use the developer portal or command-line tools, such as cURL.

Check out this example to pass a key in the request header using cURL:

```
curl --header "Ocp-Apim-Subscription-Key: <your key string>"
https://<yourapim gateway>.azure-api.net/api/path
```

Here's another example cURL command that passes a key in the URL as a query string:

```
curl https://<yourapim gateway>.azure-api.net/api/
path?subscription-key=<keystring>
```

A **401 Access Denied** response will be received from the API gateway if the key is not passed in the header for any reason, the same as a query string in the URL.

Securing APIs by using certificates

In the previous section, we used keys, but we can also use certificates to provide **Transport Layer Security (TLS)** mutual authentication between the client and the API gateway.

We have to configure the API Management gateway to only allow requests with certificates containing a specific thumbprint. The authorization at the gateway level is handled through inbound policies.

A certificate can include the following properties:

- **Certificate authority (CA)**: The API allows only certificates signed by a particular CA
- **Thumbprint**: The API allows only certificates including a specified thumbprint
- **Subject**: The API allows a specific subject (mentioned in the certificate)
- **Expiration date**: The API only allows certificates that have not expired and are still available

Accepting client certificates in the consumption tier

If you build your APIs from serverless technologies, such as Azure Functions, the Consumption tier in API Management is designed to meet serverless design principles.

You must explicitly enable the use of client certificates for the Consumption tier, which you can do on the **Custom domains** page.

A custom domain is a domain or subdomain we can buy from a provider and use for personal or professional purposes (e.g., yourdomain.com or pages.yourdomain.com).

Figure 2.14 – Custom domains

In the previous screenshot, we can see that we are able to add one or more client certificates. We can enable requesting a client certificate by selecting **Yes**. Next, we will create policies.

Certificate authorization policies

There are the following different authorization policies:

- Inbound policies are executed when the API Management API is called
- Backend policies are executed when API Management calls the backend APIs
- Outbound policies are executed when API Management returns the response to the caller

We can create policies in the **Inbound processing** policy file within the API Management gateway:

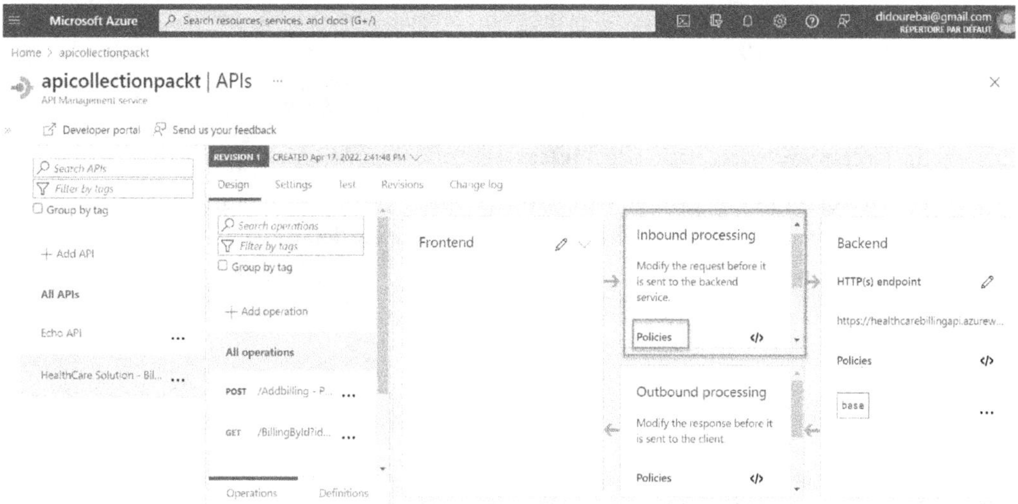

Figure 2.15 – Inbound processing | Policies

To secure an API, we are able to add a subscription and keys in Azure API Management. We can also use certificates and policies. Next, we will learn how we can create a backend API and deploy it.

Exercise 1 – creating a backend API and deploying APIs

During this exercise, you'll learn how to do the following:

- Create an API Management instance
- Import an API
- Configure the backend settings
- Test the API

We will start by creating an API Management instance using two methods, the first one using the Azure portal and the second using Azure Cloud Shell.

Creating an API Management instance

This section will show you how to create an API Management instance using two methods: the **Azure portal** and **Azure Cloud Shell**.

Using the Azure portal

We will open the Azure portal using the following link: `https://portal.azure.com/`, then, follow these steps:

1. Select **Create a resource** from the home page.

2. On the **Create a resource** page, select **Integration** and then **API Management**.

3. On the **Install API Management gateway** page, enter the different settings requested. The following screenshot shows the **Project details** and **Instance details** sections:

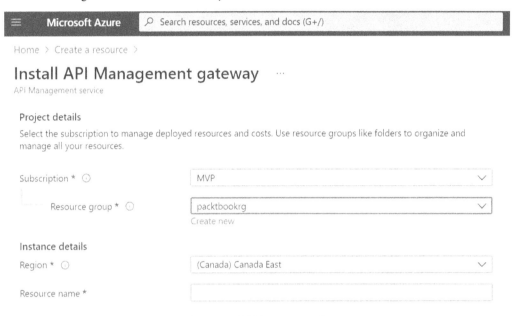

Figure 2.16 – Create an API Management instance part 1

In the **Project details** section, we need to select the **Subscription** field option and select the **Resource group** field option or create a new one.

In the **Instance details** section, we will specify the **Region** field and enter the **Resource name** field. This is a unique name for your API Management service and can't be modified after creation. A default domain name will be generated as follows: `<ResourceName>.azure-api.net`. However, we can configure a domain name to change it.

If we scroll down in the window shown in the previous screenshot, we'll see that we need to complete the **Organization name** and **Administrator email** fields, as well as select an option from the **Pricing tier** field. It will look as shown in the following screenshot:

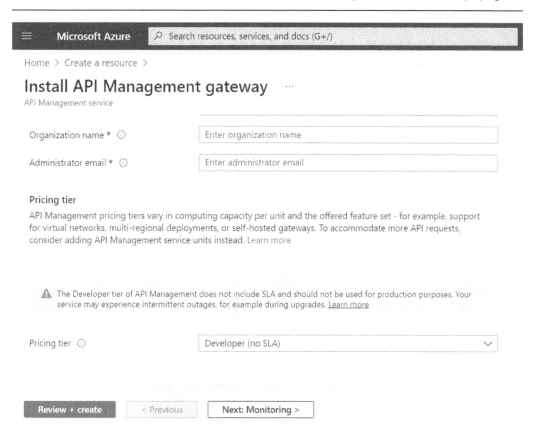

Figure 2.17 – Create an API Management instance part 2

Enter the organization name, which will be used in many places, including in the title of the developer portal and when sending notification emails.

We also need to enter the administrator email address to receive all the notifications sent from API Management.

At the end of the blade, you need to select the pricing tier. You can start by using the Developer tier to evaluate the service, but it is not recommended for production use. We can scale the API Management tiers. Note that scale units are not supported for the Developer and Consumption tiers.

Then, select **Review + create**.

> **Important note**
> Creating and enabling an API Management service in this tier can take 30-40 minutes. You can pin the service to the dashboard afterward so you can find it quickly.

We created an API Management service by using the Azure portal. It is an interactive method, but if you like using the command line, you can use Azure Cloud Shell. We will learn about that in the next section.

Using Azure Cloud Shell

Azure Cloud Shell is a browser-based interactive, authorized shell for administering Azure resources. It gives you the option of using either Bash or PowerShell as your shell experience, depending on how you operate:

1. Log in to the Azure portal and open Cloud Shell.

Figure 2.18 – Azure Cloud Shell

When the Cloud Shell portal opens, you have two options: PowerShell or Bash. We will select the **Bash** environment:

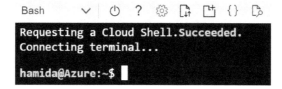

Figure 2.19 – Azure Cloud Shell using the Bash environment

2. We will define variables in order to use them in **Command-Line Interface** (**CLI**) commands. In the following code snippet, we have just defined the name of the API Management instance, the location, and the email address:

    ```
    $ myApiName=apicollectionpackt
    $ myLocation=EastUS2
    $ myEmail=rebai.hamida@gmail.com
    ```

3. We will create a resource group. The following commands will create a resource group named packtbookrg:

    ```
    $ az group create --name packtbookrg --location
    $myLocation
    ```

4. We will create an API Management instance. The `az apim create` command is used to create the instance. The `--sku-name Consumption` option is used in our case just to speed up the process for the walk-through, but we can use any **Stock-Keeping Unit** (**SKU**) (pricing tier):

```
$ az apim create -n $myApiName \
--location $myLocation \
--publisher-email $myEmail \
--resource-group packtbookrg \
--publisher-name First-APIM-Exercise \
--sku-name Consumption
```

> **Important note**
> The operation should complete in about 5 minutes.

After creating the API Management service, we will import and publish an API in the next section.

Importing an API

This section shows you how to import and publish an OpenAPI Specification backend API (formerly known as the Swagger Specification).

In this section, you'll learn about the **minimal web API** used in our sample.

Introduction to the minimal web API

This is a new approach for building APIs without all the complex structure of a **model-view-controller** (**MVC**) that is minimal according to the name. It includes the essential components needed to build HTTP APIs. All you need are `csproj` and `program.cs`.

The following are the benefits of using the Minimal Web API:

- It is less complex than using ASP.NET Web API

- It is easy to learn and use

- You don't need the MVC structure – no controllers!

- It requires minimal code to build and compile the application, which means the application runs much faster, so better performance

- It has the latest improvements and functionalities of .NET 6 and C#10

Creating a minimal web API using Visual Studio 2022

In this section, we will create an **ASP.NET Core Web API** application using Visual Studio 2022.

Follow these steps to create a minimal web API in Visual Studio 2022:

1. Open Visual Studio 2022 and select **Create a new project**.

2. Select the **ASP.NET Core Web API** project template:

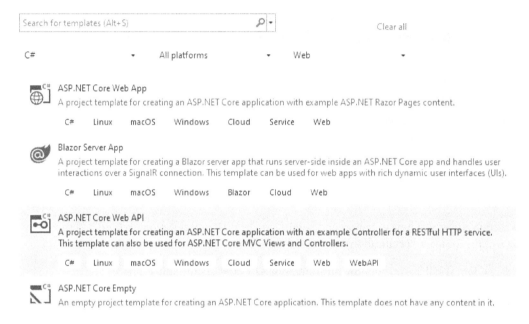

Figure 2.20 – The ASP.NET Core Web API template

3. Disable **Use controllers (uncheck to use minimal APIs)** to be able to use minimal APIs.

Figure 2.21 – Enable Minimal APIs in Visual Studio 2022

4. When the project is created, open `program.cs` to see the API implementation:

```
app.MapGet("/weatherforecast", (HttpContext httpContext) =>
{
    var forecast = Enumerable.Range(1, 5).Select(index =>
        new WeatherForecast
        {
            Date = DateTime.Now.AddDays(index),
            TemperatureC = Random.Shared.Next(-20, 55),
            Summary = summaries[Random.Shared.Next(summaries.Length)]
        })
        .ToArray();
    return forecast;
})
.WithName("GetWeatherForecast");
```

Figure 2.22 – Minimal API sample in Program.cs

After exploring the minimal web API, we will learn how we can import it into the API Management service in the next section.

Importing an API to API Management

Let's look at the steps to import an API to API Management:

1. In the Azure portal, search for and select **API Management services**.

2. Next, on the **API Management** window, select the API Management instance you have already created.

3. Select **APIs** in the API Management services navigation pane.

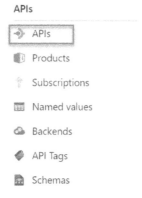

Figure 2.23 – APIs in API Management

4.   Select **OpenAPI** from the list. After that, we will select **Full** in the popup:

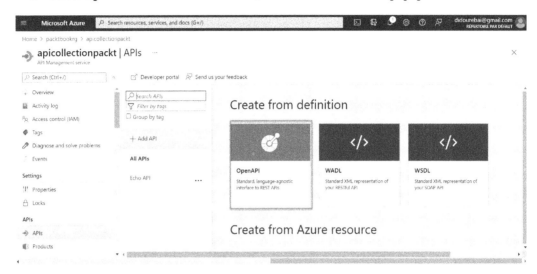

Figure 2.24 – Select OpenAPI

We need to fill in all the information needed in the **Create from OpenAPI specification** popup:

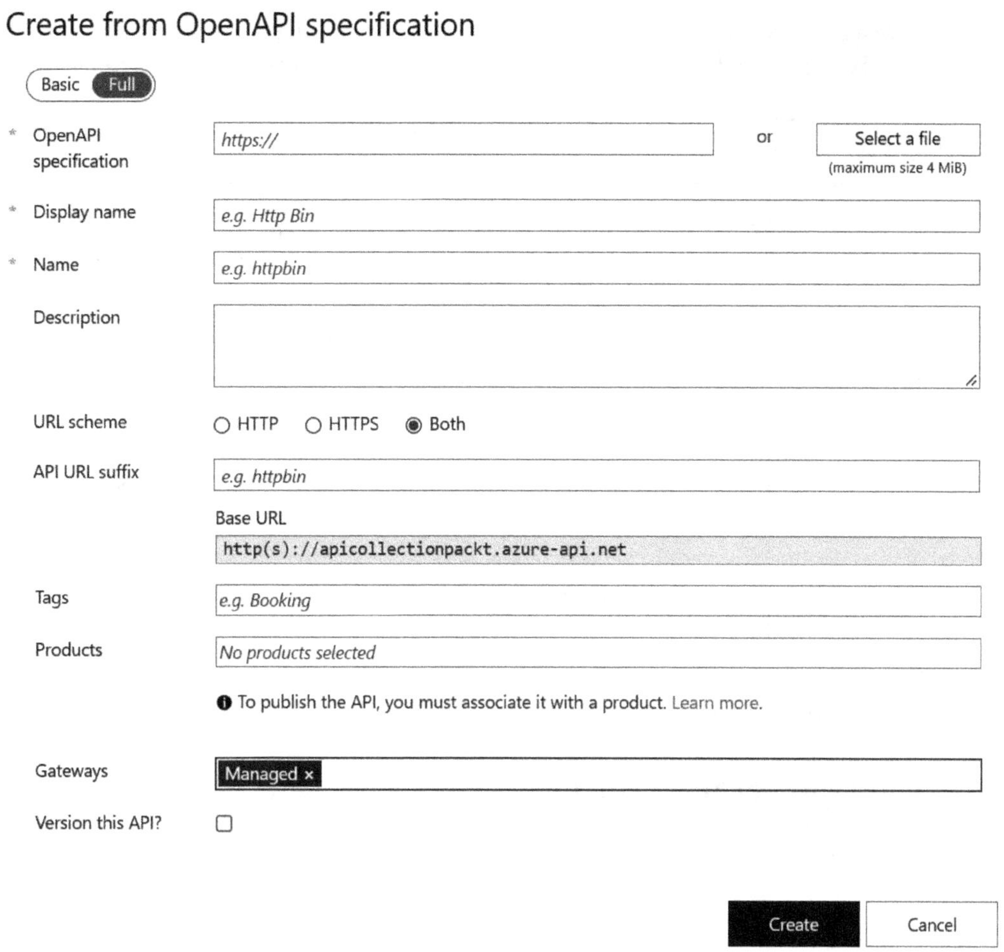

Create from OpenAPI specification

(Basic **Full**)

* OpenAPI specification	*https://*	or	**Select a file**
			(maximum size 4 MiB)

* Display name

e.g. Http Bin

* Name

e.g. httpbin

Description

URL scheme   ○ HTTP   ○ HTTPS   ⦿ Both

API URL suffix

e.g. httpbin

Base URL

`http(s)://apicollectionpackt.azure-api.net`

Tags

e.g. Booking

Products

No products selected

❶ To publish the API, you must associate it with a product. Learn more.

Gateways

[Managed ×]

Version this API?   ☐

Create   Cancel

Figure 2.25 – Create from OpenAPI specification

5. To create an API from **OpenAPI specification**, we need to add all the configuration settings. The following table describes every setting displayed in the previous screenshot:

Setting	Description
OpenAPI specification	This refers to the service implementing the API. Requests will be forwarded to this address. Once the URL is entered, most of the information needed in the form is automatically filled in.
Display name	This name will be displayed in the developer portal after configuration.
Name	This is a unique name for the API.
Description	This is optional to provide more information related to your API.
API URL suffix	API Management distinguishes APIs by their suffix. This suffix is unique for each API. It will be added to the base URL of the API Management service.

Click on **Create** to import the API.

Configuring the backend settings

The API is created, and a backend now needs to be specified.

We will select the settings in the blade to the right and enter the URL in the **Web service URL** field. After that, we will deselect the **Subscription required** checkbox.

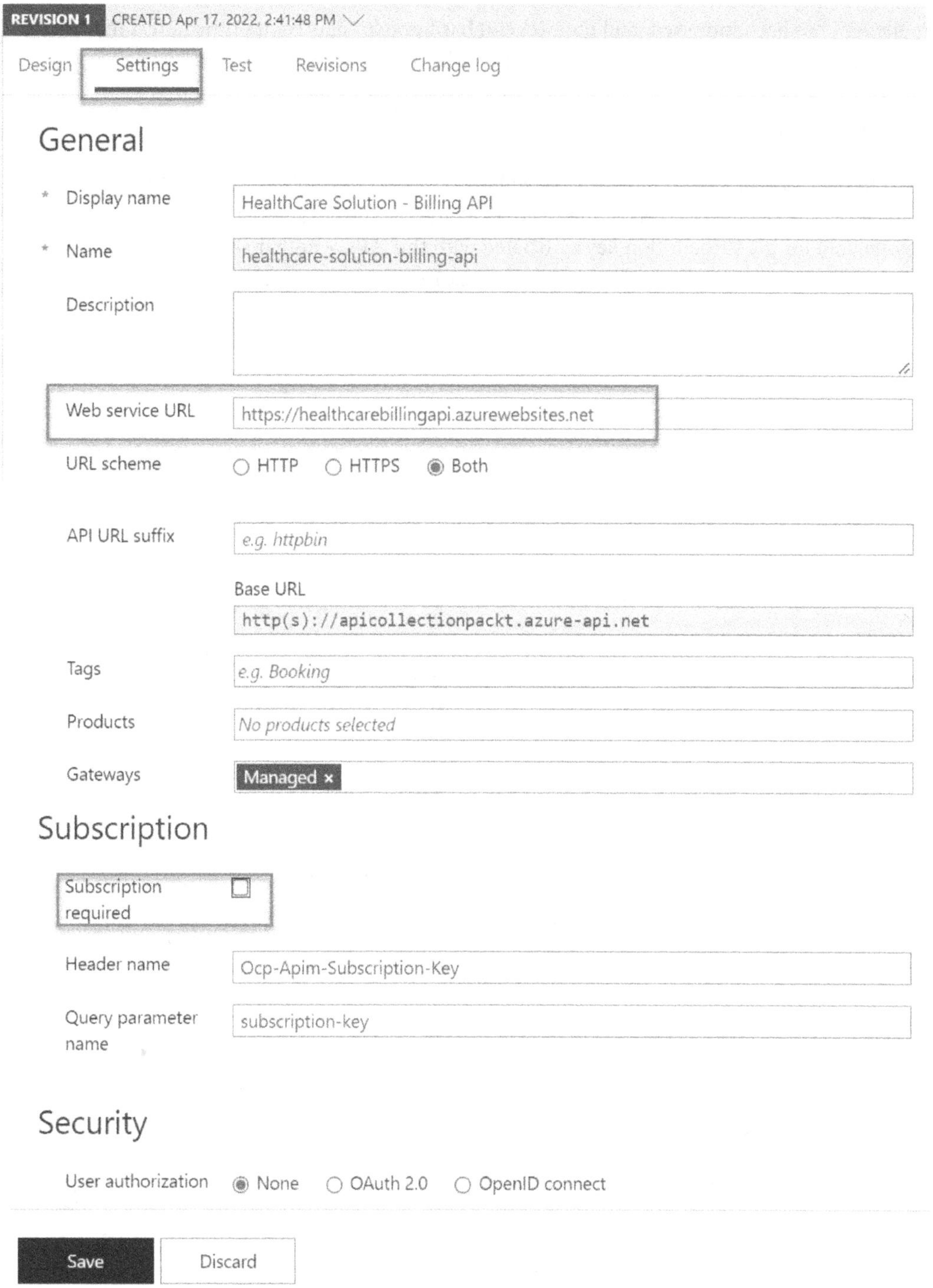

REVISION 1   CREATED Apr 17, 2022, 2:41:48 PM ∨

Design   Settings   Test   Revisions   Change log

General

*   Display name

HealthCare Solution - Billing API

*   Name

healthcare-solution-billing-api

Description

Web service URL

https://healthcarebillingapi.azurewebsites.net

URL scheme   ○ HTTP   ○ HTTPS   ◉ Both

API URL suffix

e.g. httpbin

Base URL

http(s)://apicollectionpackt.azure-api.net

Tags

e.g. Booking

Products

No products selected

Gateways

Managed ✕

Subscription

Subscription required ☐

Header name

Ocp-Apim-Subscription-Key

Query parameter name

subscription-key

Security

User authorization   ◉ None   ○ OAuth 2.0   ○ OpenID connect

Save   Discard

Figure 2.26 – Configure the backend settings

Once the API has been imported, and the backend has been configured, it's time to test the API. In the next section, we will learn how to test our API in the Azure Management API service.

Testing the API

We need to select the **Test** tab. On the left, we will find all operations related to our API. On the right of the page are the query parameters and headers, if any. `Ocp-Apim-Subscription-Key` is auto-populated for the subscription key associated with this API. Another selection will show a list of methods. We can select a method, for example, `BillingList`, then click on **Send**.

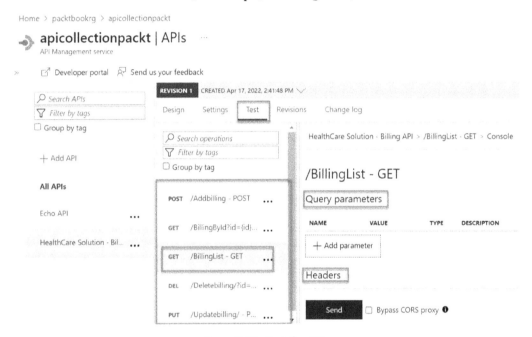

Figure 2.27 – Test the API

The backend responds with 200 OK and some data.

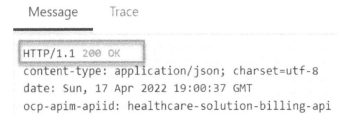

Figure 2.28 – Backend response 200 OK

In the next section, we will learn how we can use Azure API Management to proxy a public API.

Exercise 2 – using Azure API Management to proxy a public API

When we design a solution to consume one or more APIs, communication between the different services, systems, and scripts will have to go through an API. Before describing these communications and designing our APIs, we must take into consideration several elements, such as the following:

- The route structure
- Authentication and authorization
- Rate limiting

With Azure API Management, before data is received or sent, you can easily proxy an existing API and modify the input and output.

In most cases, we want to modify the structure of an existing public API, add authentication, limit the number of incoming requests, or even cache the results obtained. This is what we will cover in this section, and we will discover the ease of managing an API and consuming it quickly.

We can use an Azure Logic Apps proxy, Azure App Service, or Azure Functions apps, but in our example, we will use a public API.

In this exercise, we will use the previously imported API.

Importing an OpenAPI schema for proxying

There are different formats that the Azure API Management service will import.

In the list of available methods, the `/BillingList` method provides a list of billing details in JSON format. Using a standard PowerShell query, as seen in the following code snippet, will allow us to retrieve the results through the API proxy:

```
Result = Invoke-RestMethod -URI '< https://
healthcarebillingapi.azurewebsites.net/BillingList>'
$Result.data.summary
```

Developers tend to make changes to output keys returned to them. We are going to select the method we need to modify; thereafter, we will select the same methods in **Outbound Processing** and select **Add Policy**. After that, select **Other policies**, and in **Transformation policies**, look for the **Find and replace string in body** option.

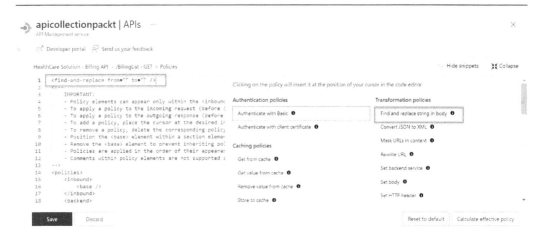

Figure 2.29 – Find and replace string in body

We will be modifying the result sent back to the client as follows:

```
find-and-replace from="price_billing" to="price" />
```

In **Outbound Processing**, we are able to modify the response before it is sent to the client.

Summary

In this chapter, we learned about the different components of the API Management service and their functions. After that, we focused on the role of API gateways in managing calls to our APIs. Then, we learned how to secure access to APIs using subscriptions and certificates. We created a backend API and used Azure API Management to proxy a public API.

In the next chapter, we will learn how to develop event-based and message-based solutions. We will explore Event Grid, Azure Event Hubs, and Azure message queues.

Developing Event-Based and Message-Based Solutions

In this chapter, we will explore the use of event-based and message-based solutions. We will publish and subscribe from a web application, see how we can send and receive messages from a Service Bus queue by using the same application, discover the operating mode of Event Grid and how it connects to services and event handlers, and explore Azure Event Hubs.

In this chapter, we're going to cover the following main topics:

- Exploring Event Grid and Azure Event Hubs
- Exercise 1 – publishing and subscribing from a .NET app to Event Grid events
- Exploring Azure message queues
- Exercise 2 – creating an Azure Service Bus namespace and a queue
- Exercise 3 – publishing messages to a Service Bus queue using a .NET Core application
- Exercise 4 – reading messages from a Service Bus queue using a .NET Core application
- Exercise 5 – sending and receiving messages to and from a topic

Introduction

An event is an action that causes a change of state. This state change can be simple or complex, but when a state changes, this is considered an event. In the event-driven architecture approach, consumers *subscribe* to events and can receive notifications when they occur. This contrasts with the traditional server-client model, in which a client actively requests updates on a set of information. This model solves a simple problem – how can we notify consumers of status changes? The message-based approach solves this same problem but in a different way than the traditional server-client model. The simple idea behind the message-based approach is that instead of consumers subscribing to event updates, individual elements or services can have message queues with many calls. These calls are ordered and sent to all parties.

To summarize, instead of having an event creating a service through a notification, a message queue takes an event or output that requires additional processing and then adds it to a long queue.

Business organizations need to become event driven and able to respond in real time. In an event-driven architecture, the events are used to trigger and communicate between decoupled services and delivered in nearly real time, and the consumers are able to respond immediately when events are produced.

An event can be a change of status or an update or deletion of an item placed in a basket on an e-commerce site. A sequence of related events represents a behavior or a status. For example, for an item purchased, its price and a delivery address, in this case, the events will present the identifiers, such as sending a notification that an order has been shipped.

Let's examine the following diagram. The event router links the different services and is used as a route for sending and receiving messages.

When an event producer is generated, the event router will execute a response to the first event after it sends it as input to the appropriate consumers. We can observe the flow of events that are handled asynchronously between managed services whose outcomes are determined as a result of a service's reaction to an event.

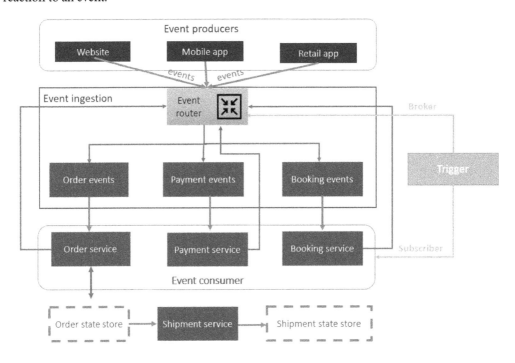

Figure 3.1 – Example of event-driven architecture

We have examined the basic elements of event-based architecture. We will now move on to look at **Azure Event Grid**, which allows us to easily build applications with event-based architecture, and **Azure Event Hub**, a scalable event-processing service that allows us to store and manage events.

Exploring Event Grid and Azure Event Hubs

In this section, we will explore the basic elements of Event Grid and Event Hub.

Event Grid

Azure Event Grid is used to easily build applications with event-based architectures.

It's simple and interactive. To start, we select the Azure resource that we want to subscribe to, and then we indicate the event handler or the endpoint of the webhook to which we will send the event.

Event Grid supports all events from Azure services, such as storage blobs and resource groups. Event Grid also supports other external events using custom topics.

The filters provide the ability to route specific events to different endpoints. We can also multicast to multiple endpoints to ensure events are reliably streamed.

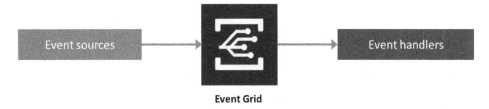

Figure 3.2 – Sources and handlers

An event source is where the event happens. Several Azure services, including Blob Storage, Media Services, IoT Hubs, and Service Bus, are automatically configured to send events. We can also use custom applications to send events. Note that custom applications don't need to be hosted in Azure or in the cloud to use Event Grid for event distribution.

An event handler is the destination the event is sent to. The handler takes some further action to process the event. Several Azure services, such as **Azure Functions**, **Logic Apps**, **Queue Storage**, or any event hubs, are automatically configured to handle events. We can also use any webhook to handle events. The webhook doesn't need to be hosted in Azure or in the cloud to handle events.

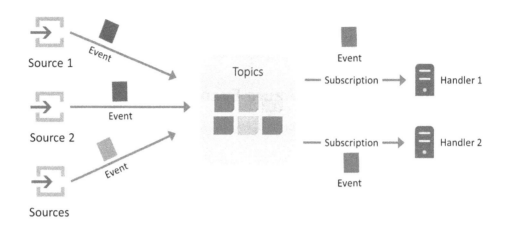

Figure 3.3 – Event Grid basic concepts

We defined Azure Event Grid as routing all the events from Azure services. An Event Grid is composed of two basic elements: the event sources and the event handlers. In the next section, we define Event Hubs before moving on to some exercises.

Event Hubs

Azure Event Hubs is a scalable event processing service that stores and processes large volumes of events, data, and telemetry produced by distributed software and devices. It provides a distributed stream-processing platform with low latency and high reliability.

Exercise 1 – publishing and subscribing from a .NET app to Event Grid events

In this exercise, we can use either Visual Studio Code or Visual Studio 2022 to create a console .NET application. We will work through the following steps:

1. Creating an Event Grid.
2. Creating a .NET Console project.
3. Making some modifications to the program class to be able to connect to Event Grid.
4. Publishing new events to the Event Grid.

We will start by creating an Event Grid topic in the next section.

Creating an Event Grid topic

An Event Grid topic is a channel for related events (for example, *storage events* or *inventory events*).

An Event Grid topic provides an endpoint to which sources send events. A publisher creates an Event Grid topic and decides whether the event source requires one topic or multiple topics. Topics are used for collections of related events. Subscribers decide which topics to subscribe to so they can respond to particular types of events.

We will use the Azure portal to create an Event Grid topic as follows:

1. Browse to the Azure portal (`https://portal.azure.com`), and then sign in with your account.
2. In the **Search services and marketplace** text box, type `Event Grid Topic` and press *Enter*.
3. Select the **Event Grid Topic** result, then select **Create**. The following screenshot displays the configured settings in the **Basics** tab. We enter all the information needed to create our Event Grid topic, then select **Review + create**, and then **Create**.

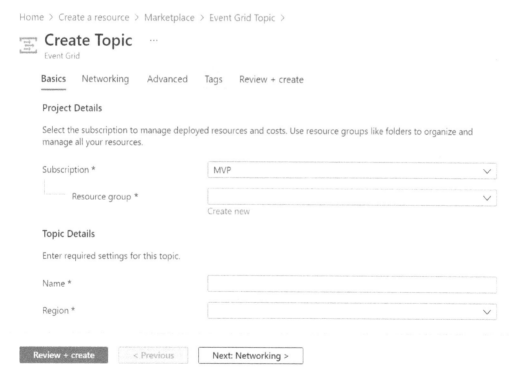

Figure 3.4 – Create Topic in Event Grid

4. We will now deploy the Azure Event Grid Viewer to a web application.

We have created an Event Grid topic. In the next section, we will create a web application in order to deploy the Azure Event Grid viewer.

Creating a web app to deploy the Azure Event Grid viewer

We will create a web app in the same resource group as our Event Grid topic. We will publish the web app instance in a Docker container using the following configuration settings:

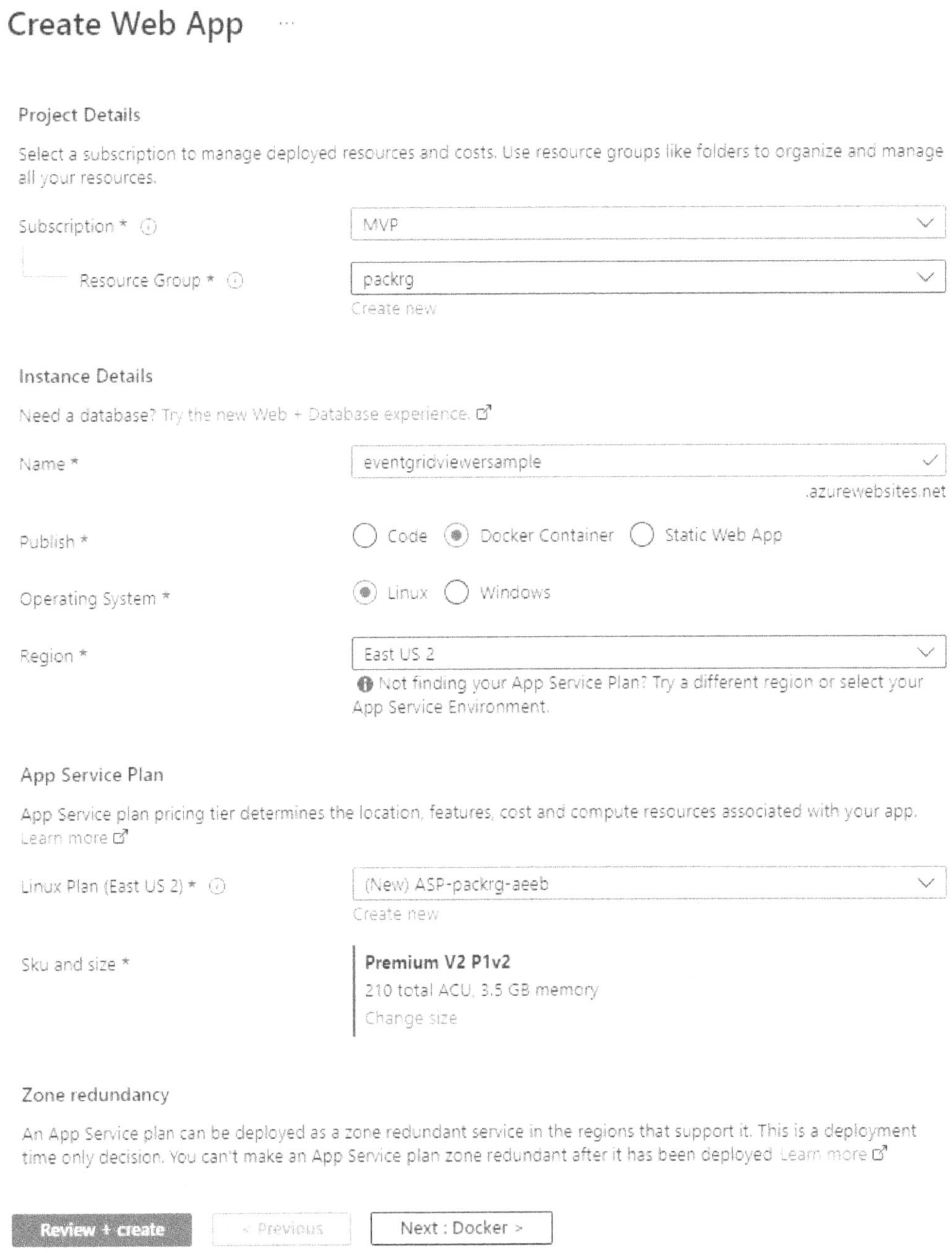

Create Web App   ...

Project Details

Select a subscription to manage deployed resources and costs. Use resource groups like folders to organize and manage all your resources.

Subscription * ⓘ   MVP

Resource Group * ⓘ   packrg

Create new

Instance Details

Need a database? Try the new Web + Database experience. ⬀

Name *   eventgridviewersample

.azurewebsites.net

Publish *   ○ Code   ● Docker Container   ○ Static Web App

Operating System *   ● Linux   ○ Windows

Region *   East US 2

ⓘ Not finding your App Service Plan? Try a different region or select your App Service Environment.

App Service Plan

App Service plan pricing tier determines the location, features, cost and compute resources associated with your app. Learn more ⬀

Linux Plan (East US 2) * ⓘ   (New) ASP-packrg-aeeb

Create new

Sku and size *   **Premium V2 P1v2**
210 total ACU, 3.5 GB memory
Change size

Zone redundancy

An App Service plan can be deployed as a zone redundant service in the regions that support it. This is a deployment time only decision. You can't make an App Service plan zone redundant after it has been deployed. Learn more ⬀

[Review + create]   [< Previous]   [Next : Docker >]

Figure 3.5 – Create Web App – the Basic tab

In the **Docker** tab, configure the settings as shown in the following figure. We use `microsoftlearning/ azure-event-grid-viewer:latest` in the **Image and tag** field. This image is a web application built using **ASP.NET Core** and **SignalR** created for the purpose of viewing notifications from Azure Event Grid in near-real time. After this is complete, select **Review + create**.

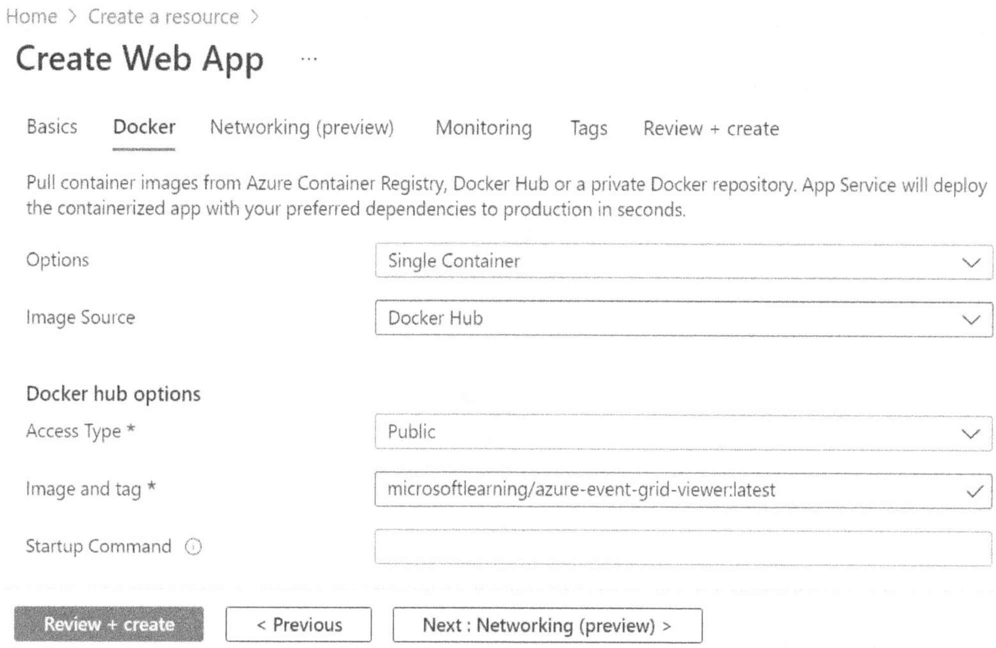

Figure 3.6 – Create Web App – the Docker tab

Lastly, select **Create** to create the web app using your specified configuration.

Creating an Event Grid subscription

We will now create a new subscription, validate its registration, then save the necessary credentials in order to publish a new event on the subject.

If we go back to the Event Grid topic we created previously, we will see at the bottom a button for creating a subscription, as shown in the following screenshot:

Figure 3.7 – Event Grid Topic window

Select the **Create one** button.

In the **Basic** tab, we start by defining the name of the event subscription and the event schema. The topic is already defined for us. We can add the event type to the destination. The following screenshot presents the configuration settings in full:

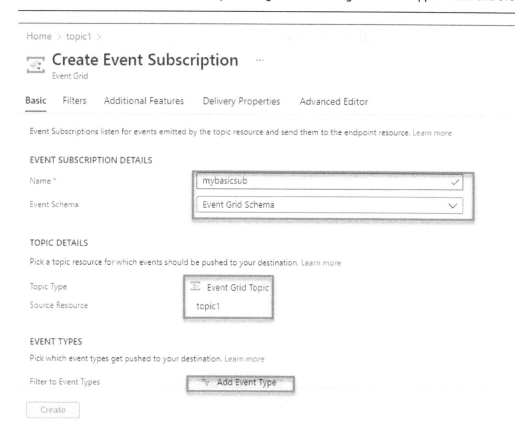

Figure 3.8 – Create Event Subscription – the Basic tab part 1

In the **ENDPOINT DETAILS** section, we need to define the **Endpoint type** field as **Web Hook**. In the **Subscriber Endpoint** text box, enter the web app URL value that we created before and add `/api/updates`. So, for example, if your web app URL value is `https://eventgridviewersample.azurewebsites.net`, then your **Subscriber Endpoint** value would be `https://eventgridviewersample.azurewebsites.net/api/updates`. Finally, we confirm the selection and select **Create**.

If you need more details about Event Grid, you can go to the following links:

- `https://learn.microsoft.com/en-us/azure/event-grid/create-view-manage-system-topics`

- `https://learn.microsoft.com/en-us/azure/event-grid/custom-event-quickstart`

- `https://learn.microsoft.com/en-us/azure/event-grid/scripts/event-grid-cli-subscribe-custom-topic`

The following screenshot presents the endpoint details in the **Basic** tab:

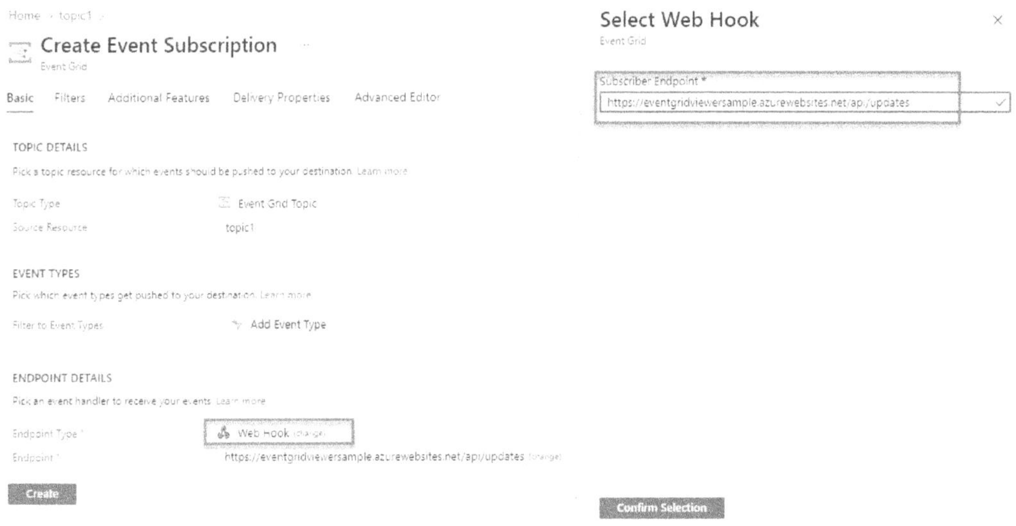

Figure 3. 9 – Create Event Subscription – the Basic tab part 2

So far, we have created the Event Grid topic and the subscription. Next, we will learn how to publish new events to the Event Grid topic using a .NET console application.

Create a .NET Console project

To create a new .NET Console project, we can use either of two **integrated development environments (IDEs)**: Visual Studio Code or Visual Studio 2022.

Using Visual Studio Code

In Visual Studio Code, open the integrated terminal. Next, run the following command to create a new .NET project named EventPublisherSample in the current folder:

```
dotnet new console -name EventPublisherSample -output .
```

Let's import the latest version of Azure.Messaging.EventGrid from NuGet, in our example, version 4.10.0, and run the following command:

```
dotnet add package Azure.Messaging.EventGrid -version 4.10.0
```

To build the application, we use the following command:

```
dotnet build
```

Using Visual Studio 2022

We select **Create a new project**, then search for the console template. We name our project EventPublisherSample and select the latest .NET framework:

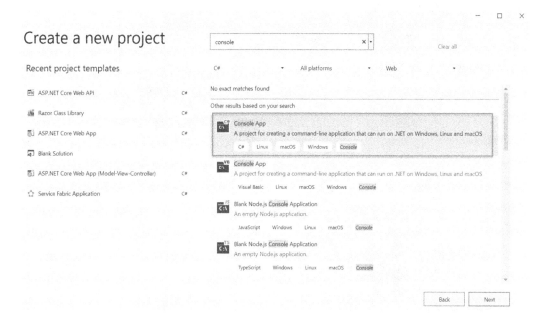

Figure 3. 10 – Create a new project – Console App

Let's import the latest version of Azure.Messaging.EventGrid from NuGet. In our example, it is version 4.10.0. We use NuGet by right-clicking on **Dependencies**, then selecting **Manage NuGet Packages,** and installing Azure.Messaging.EventGrid via NuGet, as in the following screenshot:

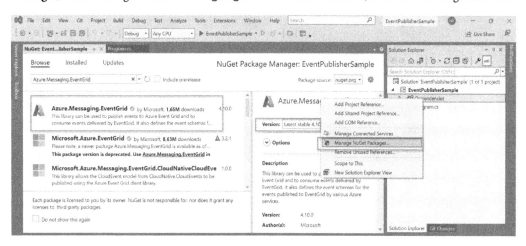

Figure 3.11 – Import Azure.Messenging.EventGrid

After adding the package, we will update our Program class in the next section.

Making some modifications to the Program class to be able to connect to Event Grid

We will continue to use Visual Studio 2022 (we can use Visual Studio Code also) in the following steps:

1. Open the `Program.cs` file.

2. Import the `Azure` and `Azure.Messaging.EventGrid` namespaces (`https://www.nuget.org/packages/Azure.Messaging.EventGrid/`) from the `Azure.Messaging.EventGrid` package imported previously from NuGet using the following lines of code:

    ```
    using Azure;
    using Azure.Messaging.EventGrid;
    ```

3. Check the entire source code in the `Program.cs` file:

    ```
    Using System;
    using System.Threading.Tasks;
    using Azure;
    using Azure.Messaging.EventGrid;
     public class Program
     {
         private const string topicEndpoint = "<topic-
    endpoint>";
         private const string topicKey = "<topic-key>";
         public static async Task Main(string[] args)
         {
         }
     }
    ```

We added the `System` and `System.Threading.Tasks` packages that will be used for asynchronous methods.

In the code, we added a `Program` class that includes two private properties and a static asynchrony method.

The first property is the `topicEndpoint` string constant that includes the Event Grid endpoint value. The second property is the `topicKey` string constant that contains the Event Grid topic key value.

We can find these credentials by going to the **Event Grid Topic** blade, selecting **Overview** in the left-hand menu, and looking at the **Topic Endpoint** entry:

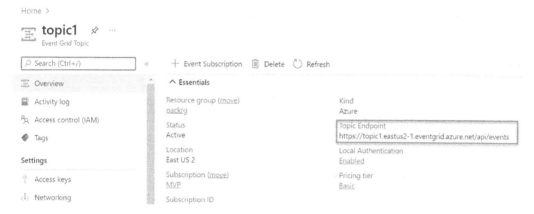

Figure 3.12 – Topic Endpoint

The topic key is obtained by selecting **Access keys** under the **Settings** heading in the same left-hand menu as previously. We need to select and copy the link in the **Key 1** text box.

Figure 3.13 – Topic key

In the next section, we will publish new events.

Publishing new events

In the Program.cs file, in the empty main method, we will create a new variable named credential that will use topicKey, and another variable named client that will use the endpoint and the credential variables in the constructor as parameters. So, we will add the following lines of code to the main method:

```
var endpoint = new Uri(topicEndpoint);
var credential = new AzureKeyCredential(topicKey);
```

```
    var client = new EventGridPublisherClient(endpoint,
  credential);
```

We add the following source code to create a new `event1` variable and populate that variable with sample data:

```
var event1= new EventGridEvent(
    subject: $"New Patient: Hamida Rebai",
    eventType: "Patients.Registration.New",
    dataVersion: "1.0",
    data: new
    {
        FullName = "Hamida Rebai",
        Address = "Quebec, G2C0L6"
    }
  );
```

Next, we add an `event2` variable:

```
var event2 = new EventGridEvent(
    subject: $"New Patient: Mohamed Trabelsi",
    eventType: "Patients.Registration.New",
    dataVersion: "1.0",
    data: new
    {
        FullName = "Mohamed Trabelsi",
        Address = "Quebec, G1X3Z2"
    }
  );
```

After, we will add the following lines of code to invoke the `EventGridPublisherClient`. SendEventAsync method, and we will see in the following source code how to render the message for every event to the console:

```
        await client.SendEventAsync(event1);
        Console.WriteLine("This is the first event published");
        await client.SendEventAsync(event2);
        Console.WriteLine("This is the second event
  published");
```

Let's run our application and observe the result in the Azure Event Grid viewer web application:

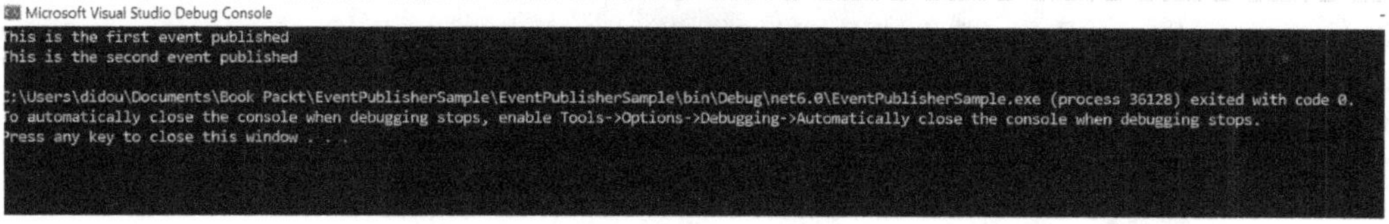

Figure 3.14 – Console App output for published events

In the following screenshot, we present **Azure Event Grid Viewer** published in Azure App Service:

Figure 15 – Azure Event Grid viewer result

To summarize this exercise, we used static data in the console application, but in real life, this data can be provided from another application (web or mobile) or any **software as a service** (**SaaS**) solution. For our use case, this data comes from a web and mobile application, and every new registration will be sent to Event Grid to be validated by another system.

Exploring Azure message queues

Microsoft Azure Service Bus is a fully managed enterprise integration message broker. Some common messaging scenarios are included in Service Bus, such as messaging in case of transferring data or decoupling applications to improve their reliability and their scalability and message sessions by implementing workflows that require message ordering, for example, or topics and subscriptions.

Azure supports the following two types of queuing mechanisms:

- Queue Storage is a simple message queue service for storing large numbers of messages
- Service Bus queues are part of a broader set of messaging services that support queuing, publishing/subscribing, and advanced integration patterns

Azure Queue Storage is a service for sending and receiving messages but is also used to store a large number of messages; a queue message can be up to 64 KB in size.

Exercise 2 – creating an Azure Service Bus namespace and a queue

In this exercise, we will perform the following actions:

- Create an Azure Service Bus namespace
- Create an Azure Service Bus queue

The queues can be created using the interactive interface of the Azure portal, using the command line via PowerShell or Azure CLI, or using Resource Manager templates. We will look at only the first two methods here: Azure Portal and Azure CLI.

Using the Azure portal

Let's open the Azure portal and select **All services**. In the **Integration** category, we can find the **Service Bus** resource:

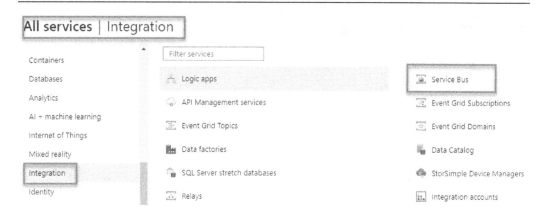

Figure 3.16 – Creating a Service Bus using the Azure portal

After filling in the configuration required to create a namespace that will include after the queue, we click on + **Create** or select **Create service bus namespace**:

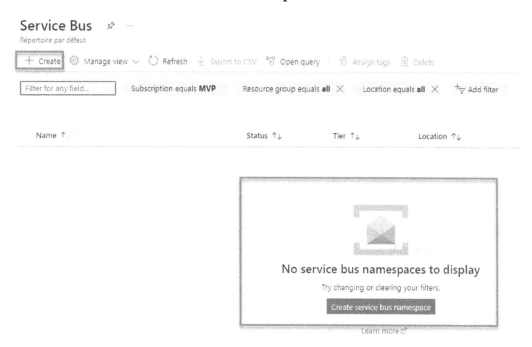

Figure 3.17 – Creating a namespace using the Azure portal

In the **Basics** tab, we have the following interface:

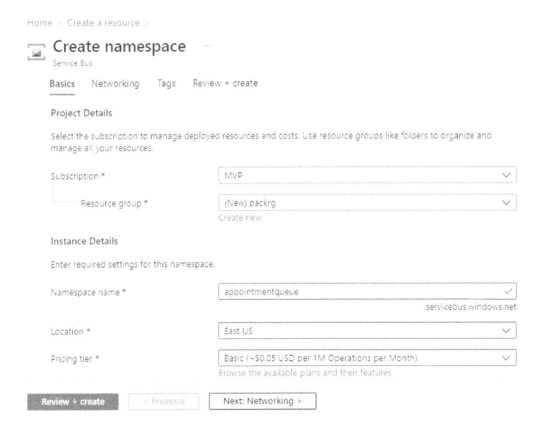

Figure 3. 18 – Creating a Service Bus namespace

If you select **Basic** in the **Pricing tier** field, note that you will not be able to create **Topics**, only **Queues**, as presented in the following screenshot:

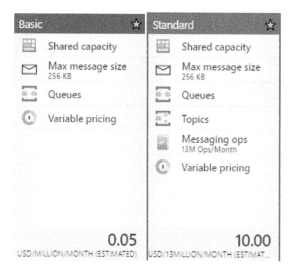

Figure 3.19 – Pricing tier plans and features – Basic and Standard

In the **Networking** tab, you can leave the default values for the connectivity method that is public access, or you can select your private access if you have already configured the landing zone. In the landing zone, you configure your account, security governance, networking, and identity.

We will select **Review + create** then select **Create** and the Service Bus namespace will be created.

Once finished, we will create a queue in the portal. Let's open the Service Bus namespace and select **Overview** from the left-hand menu. Next, at the top of the screen, select + **Queue** and fill in all the information required to create the queue, including the name, the queue max size, the max delivery count, message time to live (days, hours, minutes and seconds), the lock duration (days, hours, minutes and seconds). Additionally, we need to enable the following options:

- Auto-delete on idle queue
- Duplicate detection
- Dead lettering on message expiration
- Partitioning
- Sessions

We can enable forwarding messages to queues and topics:

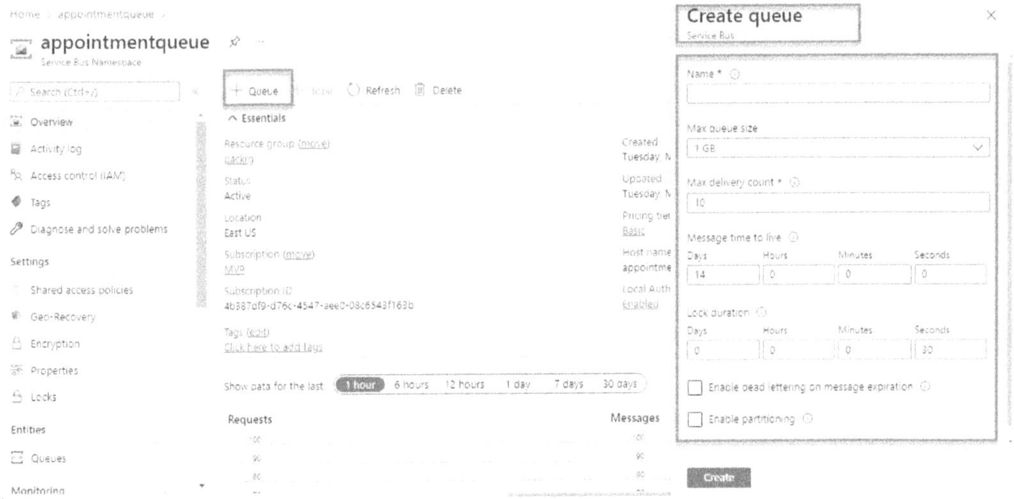

Figure 3.20 – Creating a queue

We used the Azure portal to create an Azure Service Bus namespace and a queue. In the next section, we will see how to do this using the Azure CLI.

Using the Azure CLI

To begin, we need to create a resource group if one doesn't exist already:

```
Az group create -name packrg -location eastus
```

A message queue uses a namespace, so we will need to create a namespace:

```
Az servicebus namespace create -resource-group
 packrg -name appointmentQueue -location eastus
```

Next, we create a queue using the following command:

```
az servicebus queue create -resource-group packrg -namespace-
name appointmentQueue -name myqueue
```

To check whether the message queue has been created successfully, you can check in the portal or run the following command line:

```
az servicebus queue list -resource-group packrg -namespace-name
appointmentQueue
```

In the following screenshot, we can see the different Service Bus queues:

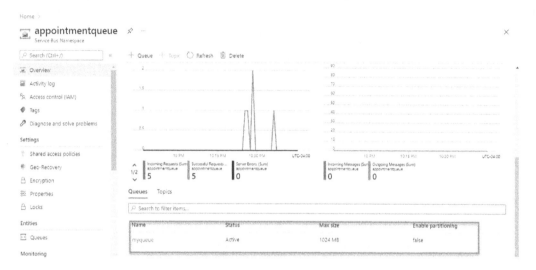

Figure 3.21 – Service Bus queues

We demonstrated two different methods to create an Azure Service Bus namespace and a queue – the Azure CLI and the Azure portal. In the next exercise, we will publish messages to a Service Bus queue using an application.

Exercise 3 – publishing messages to a Service Bus queue using a .NET Core application

We previously created the Azure Service Bus queue. Now, we will start sending messages from the Azure Service Bus queue using a .NET Core console application.

We will create a solution that will include two projects:

- `AzureQueueSample.Domains`: A class library
- `AzureQueueSample.Sender`: A .NET Core console application

We will add the `AzureQueueSample.Domains` project as a project reference to the `AzureQueueSample.Sender` project. To do that, right-click on the `AzureQueueSample.Sender` project **Dependencies**, then select **Add Project Reference**:

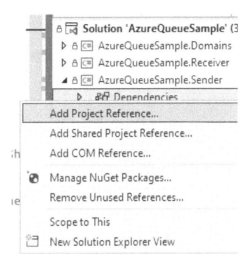

Figure 3.22 – Add Dependencies reference to a project

We select the `AzureQueueSample.Domains` project and the project will be added in as a reference:

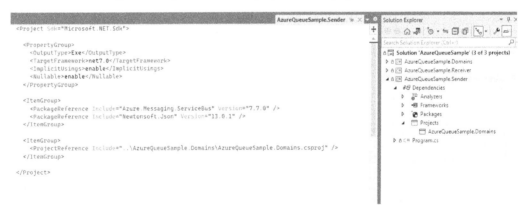

Figure 3.23 – AzureQueueSample solution

In the `AzureQueueSample.Domains` project, we will add a new class named `Appointment`:

```
Public class Appointment
    {
        public Guid AppointmentId { get; set; }
```

```
        public string Status { get; set; }
        public string SlotId { get; set; }
        public string PatientEmail { get; set; }
        public string DoctorId { get; set; }
        public string CurrentDate { get; set; }
        public string StartTime { get; set; }
        public string EndTime { get; set; }
    }
```

In the following screenshot, we use **NuGet Package Manager** to add two different packages to AzureQueueSample.Sender, namely Azure.Messaging.ServiceBus and NewtonSoft. Json.

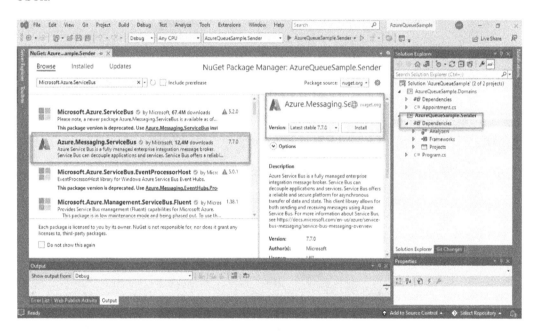

Figure 3.24 – Add Microsoft.Azure.ServiceBus package from NuGet

Next, we open the **program.cs** class of AzureQueueSample.Sender and add a list of appointments. We also add the connection string to the target Azure Service Bus namespace and the queue name.

In the Azure portal, go back to the Service Bus namespace and select **Settings | Shared access policies**, then copy the **Primary Connection String** value:

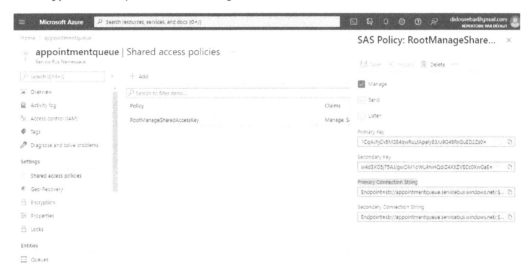

Figure 3.25 – Shared access policies in the Azure Service Bus namespace

This is the Program class that allows us to publish messages to the Service Bus queue:

```
private const string ServiceBusConnectionString = "Endpoint=sb:
//appointmentqueue.servicebus.windowsR.net/;SharedAccessKeyName
=RootManageSharedAccessKey;SharedAccessKey=1CqAJhjCv5M284qwRuLt
Apely83Ju9G48RxGuED2Zs0=";

    private const string QueueName = "myqueue";
    static List<Appointment> appointments = new
List<Appointment>
    {
        new Appointment ()
        {  AppointmentId =  Guid.NewGuid(),
           CurrentDate = DateTime.Now.ToString(),
           StartTime = "10:00AM",
           EndTime ="11:00AM",
           DoctorId = "code1",
           SlotId="S1",
           PatientEmail= "patient1@email.com",
           Status = "Pending"
        },
```

```csharp
        new Appointment()
        {
            AppointmentId =  Guid.NewGuid(),
            CurrentDate = DateTime.Now.ToString(),
            StartTime = "1:00PM",
            EndTime ="2:00APM",
            DoctorId = "code8",
            SlotId="S7",
            PatientEmail= "patient4@email.com",
            Status = "Pending"
        },
        new Appointment ()
        {
            AppointmentId =  Guid.NewGuid(),
            CurrentDate = DateTime.Now.ToString(),
            StartTime = "8:00AM",
            EndTime ="9:00AM",
            DoctorId = "code11",
            SlotId="S21",
            PatientEmail= "patient31@email.com",
            Status = "Pending"
        }
    };
    static async Task Main(string[] args)
    {
            // Because ServiceBusClient implements
IasyncDisposable, we'll create it
            // with "await using" so that it is automatically
disposed for us.
            Await using var client = new
ServiceBusClient(ServiceBusConnectionString);
            // The sender is responsible for publishing
messages to the queue.
            ServiceBusSender sender = client.
CreateSender(QueueName);
            foreach (var item in appointments)
            {
```

```
                var messageBody = JsonConvert.
    SerializeObject(item);
                ServiceBusMessage message = new
    ServiceBusMessage(Encoding.UTF8.GetBytes(messageBody));
                await sender.SendMessageAsync(message);
                Console.WriteLine($"Sending Message : {item.
    AppointmentId.ToString()} ");
            }
        Console.Read();
    }
```

The result will be as follows:

Figure 3.26 – Sending messages to the Azure Service Bus queue

In this exercise, we published messages to a Service Bus queue using a .NET Core application. In the next section, we will read the messages from a Service Bus queue using the same application.

Exercise 4 – reading messages from a Service Bus queue using a .NET Core application

Previously, we sent some messages to a Service Bus queue. In this exercise, we will read them. To do that, we will add a new AzureQueueSample.Receiver .NET Core console application. We will need to add the same NuGet packages as previously for the sender application: Azure.Messaging. ServiceBus and NewtonSoft.Json. We will also add AzureQueueSample.Domains as a reference.

To read a message from the Service Bus queue, we use the following code:

```
public class Program
{
    private const string ServiceBusConnectionString = "Endpoint
=sb://appointmentqueue.servicebus.windows.net/;SharedAccessKey
Name=RootManageSharedAccessKey;SharedAccessKey
=1CqAJhjCv5M284qwRuLtApely83Ju9G48RxGuED2Zs0=";
    private const string QueueName = "myqueue";
    static async Task Main(string[] args)
    {
        await ReceiveMessagesAsync();
    }

    private static async Task ReceiveMessagesAsync()
    {
        await using var client = new
ServiceBusClient(ServiceBusConnectionString);
        // The receiver is responsible for reading messages
from the queue.
        ServiceBusReceiver receiver = client.
CreateReceiver(QueueName);
        ServiceBusReceivedMessage receivedMessage = await
receiver.ReceiveMessageAsync();

        string body = receivedMessage.Body.ToString();
        Console.WriteLine(body);
        Console.Read();
    }
}
```

We have now published messages to a Service Bus queue and then read the messages from the queue using a .NET Core console application in the previous source code.

We will send messages to Azure Service Bus, and we will open the Azure Service Bus instance. In the **Overview** page, we check the metrics related to the number of requests and messages, as presented in the following screenshot:

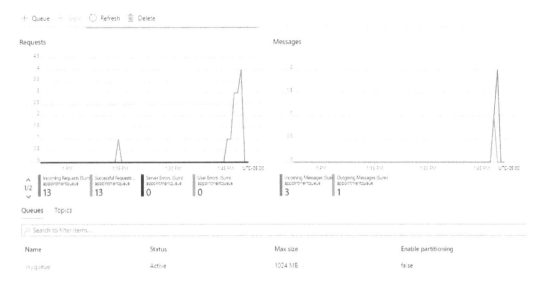

Figure 3.27 – Queue metrics in Azure Service Bus

We can select **Queues** under **Entities** to get more details related to queue metrics, as presented in the following screenshot:

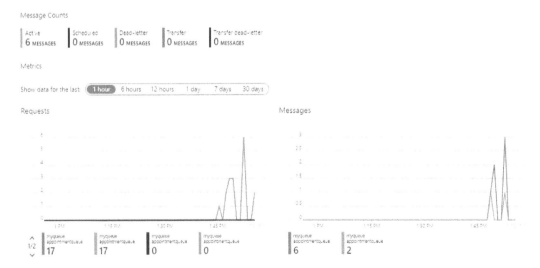

Figure 3.28 – Extended queue metrics in Azure Service Bus

In the previous sample, we sent the messages to a queue but we can also send messages to the topic. If we need to send a message in the one-to-one system, we use Azure Service Bus queue, but if we need to send a message to multiple systems, we use Azure Service Bus topic.

In the next section, we will create a topic and a subscription, then send messages to the topic.

Exercise 5 – sending and receiving messages to and from a topic

In this section, we will create a topic and a Service Bus subscription to that topic using the Azure portal. Then, we will implement a **.NET console application** to send a set of messages to this topic and receive them from the subscription.

Creating a topic using the Azure portal

To create a topic in Azure Service Bus using the Azure portal, we follow these steps:

1. Open your Service Bus namespace page and select **Topics** from the left-hand menu.

2. Click the **+ Topic** button to add a new topic as presented in the following screenshot:

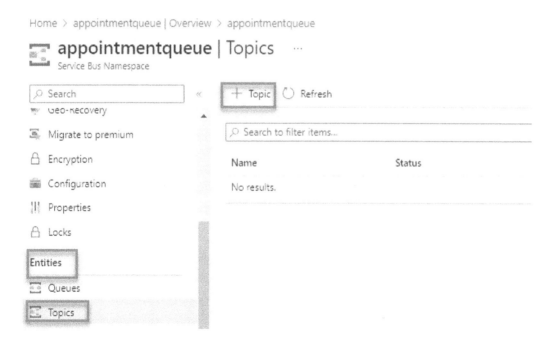

Figure 3.29 – Add a topic to your Service Bus namespace

3. A new dialog window will be displayed. Enter a name for the topic, such as `mytopic`, and leave the default values for the other options.

4. Select **Create** to confirm the topic creation.

Now that we've created a new topic, let's create a subscription to that topic.

Creating a subscription to the topic

To create a subscription to the previously created topic, we follow these steps:

1. Open the topic and click the + **Subscription** button on the toolbar, as presented in the following screenshot:

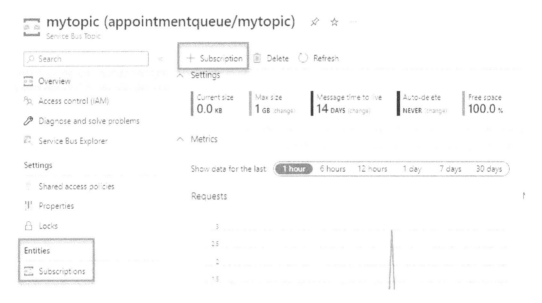

Figure 3.30 – Add a subscription to the Service Bus topic

2. A new dialog window will be displayed. To create a subscription, enter `Sub1` in the **Name** field and 3 for **Max delivery count**. We leave the other values at their defaults, as presented in the following screenshot:

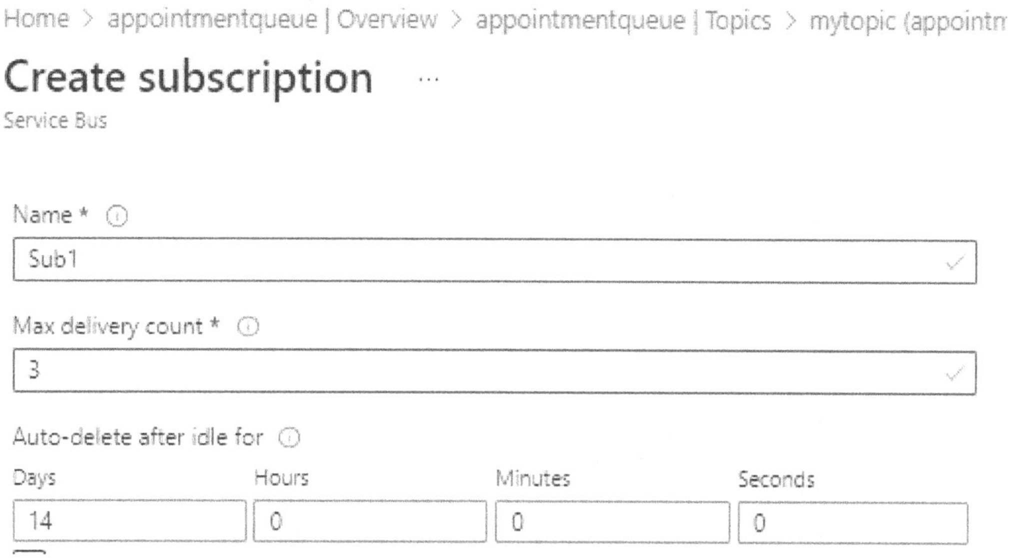

Home > appointmentqueue | Overview > appointmentqueue | Topics > mytopic (appointn

Create subscription   ...
Service Bus

Name * ⓘ

| Sub1 | ✓ |

Max delivery count * ⓘ

| 3 | ✓ |

Auto-delete after idle for ⓘ

Days	Hours	Minutes	Seconds
14	0	0	0

Figure 3.31 – Create a subscription page

3. Select **Create** to confirm the creation of the subscription.

Sending messages to the topic

In this section, we will add functions to our previously created .NET application and will send messages to the Service Bus topic.

We start by configuring the access policies for this Service Bus topic:

1. Select **Shared access policies** from the **Settings** section of the left-hand menu. Then, click the + **Add** button to configure a new shared access policy.

2. Fill in the **Policy name** field as desired and select the different policies to be authorized (**Manage**, **Send**, or **Listen**), as presented in the following screenshot:

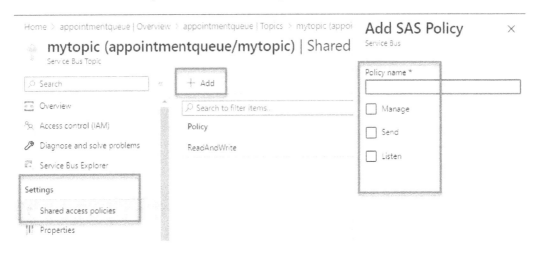

Figure 3.32 – Manage Shared access policies

3. Under **Shared access policies**, select the policy and copy the **Primary Connection String** value, as presented in the following screenshot:

Figure 3.33 – Select the Primary Connection String for the shared access policy

4. To send a message to the Service Bus topic, we use the following code:

```
private const string ServiceBusTopicConnectionString =
"Endpoint=sb://appointmentqueue.servicebus.windows.
net/;SharedAccessKeyName=ReadAndWrite;SharedAccessKey
=Fotiiqahw46U5iJY0+olkYjydpkhOrX4bfUP6NirTgY=;EntityPath=
mytopic";
    private const string topicName = "mytopic";
await using var clientForTopic = new
ServiceBusClient(ServiceBusTopicConnectionString);
        ServiceBusSender sender = clientForTopic.
CreateSender(topicName); await using var client = new
ServiceBusClient(ServiceBusConnectionString);


        ServiceBusSender sender = client.
CreateSender(topicName);
        foreach (var item in appointments)
        {
            var messageBody = JsonConvert.
SerializeObject(item);
            ServiceBusMessage message = new
ServiceBusMessage(Encoding.UTF8.GetBytes(messageBody));
            await sender.SendMessageAsync(message);

            Console.WriteLine($"Sending Message: {item.
AppointmentId} ");


        }
```

5. Open the Service Bus topic in the Azure portal, then select the subscription to see the metrics for messages sent, as presented in the following screenshot:

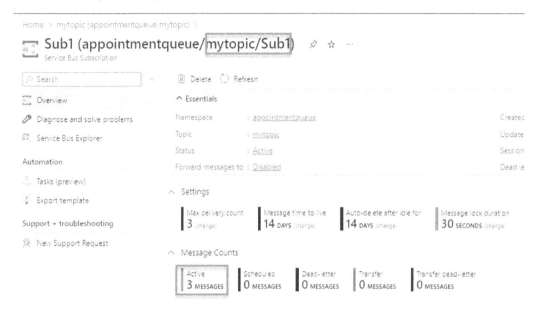

Figure 3.34 – Metrics for the Service Bus topic under subscription

6. To see more details about the requests and the messages sent to the topic, select the topic, and in the **Overview** pane, you will see something similar to the following screenshot:

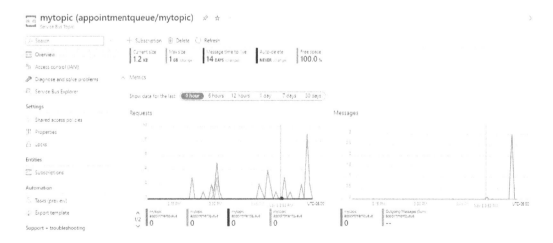

Figure 3.35 – Topic metrics: requests and messages

A queue allows a single consumer to process messages. Unlike queues, topics and subscriptions provide a one-to-many form of communication in a publish-subscribe model. This is useful when scaling to a large number of recipients. All published messages are available to all subscribed subscribers to the topic. When an editor posts a message on a topic, one or more subscribers receive a copy of the message.

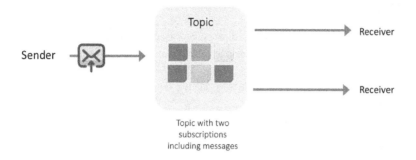

Figure 3.36 – Topics and subscriptions

Subscriptions can use additional filters to limit the messages they receive. Publishers send messages to topics the same way they send messages to queues. However, consumers do not receive messages directly from topics. Instead, the consumer receives messages from the subscription to the topic. A topic subscription is like a virtual queue that receives copies of messages sent to the topic. Consumers receive messages from subscriptions the same way they receive messages from queues.

We created a topic and a Service Bus subscription to that topic, we sent a message to the topic, and in the next section, we will receive these messages from the topic's subscription using C# source.

Receiving messages from a subscription

To read the previously created messages from the Service Bus topic, we start by creating a receiver that we can use to receive the message:

```
await using var clientForTopic = new
ServiceBusClient(ServiceBusTopicConnectionString);
        ServiceBusReceiver receiver = clientForTopic.
CreateReceiver(topicName);
The received is a different type as it contains some
service set properties:          ServiceBusReceivedMessage
receivedMessage = await receiver.ReceiveMessageAsync();
```

We will retrieve the message body as a string:

```
        string body = receivedMessage.Body.ToString();
        Console.WriteLine(body);
        Console.Read();
```

Queues and topics are similar to when the sender sends a message, but the message is processed differently, depending on the receiver. A queue can only have one consumer, but a topic can have multiple subscribers.

Summary

In this chapter, we talked about event-based and message-based solutions and the difference between them. Next, we explored Azure Event Grid and Azure Event Hubs and examined how to publish and subscribe from a .NET application to Event Grid. Then, we talked about Azure message queues and published messages from a .NET application. Lastly, we read those messages using the .NET console application.

In the next chapter, we will create and deploy function apps in Azure.

Question

1.   Which packages need to be added to read messages from a Service Bus queue using a .NET Core application?

Part 2:
Connecting Your Application with Azure Databases

In this part of the book, we will focus on Azure database solutions for relational and non-relational databases, big databases, and different types of storage.

This part comprises the following chapters:

4

Creating and Deploying a Function App in Azure

One of the critical decisions architects and developers make in the phase of architecting a new cloud software or application is how to connect to the backend services, run background processing, run backend tasks, and carry out more tasks, such as scheduling and sending emails, without affecting the main application processing. For this, the **Azure Functions** app can be useful.

We can use Azure Functions to execute code in a cloud environment in a serverless way. All we need to do is to write less code with a low cost for a specific problem without caring about the whole application, even the infrastructure where we will run it. We will focus, in this case, on logic and business scope. We can execute Azure Functions in response to events as well.

In this chapter, we will cover the basic concepts of Azure Functions and the hosting plan options. We will explore the development of Azure functions and develop durable functions, which are an extension of Azure Functions.

In this chapter, we're going to cover the following main topics:

- Exploring Azure Functions
- Developing Azure functions
- Developing durable functions

Exploring Azure Functions

Azure Functions is similar to **Azure WebJobs**, with some differences related to scaling policies, language support, and trigger events. They are both built on Azure App Service. Because Azure Functions is built on the WebJobs SDK, it shares the same triggers and connections with or to Azure services.

Azure Functions acts as a modern serverless architecture providing event-driven, configured cloud computing for application development.

In Azure Functions, there are two very important concepts: bindings and triggers.

When we create Azure Functions, we need to configure the name of the function. To call a function, we use triggers. A function must have at least one trigger. The function name and the triggers are configured in the function.json file. We have more information in the configuration file, such as on bindings. We use constraints to have input and output bindings, and all of them will allow you to connect with databases from your code. All of this saves coding on the connection. But with many Azure service pools, you don't need to do this—you don't need to write code because the bindings take care of those connections.

In the following figure, we present the different bindings and triggers:

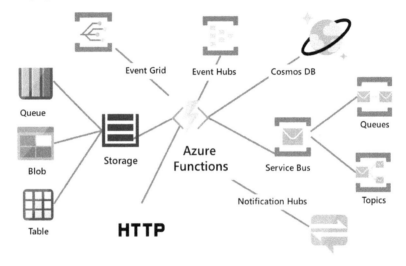

Figure 4.1 – Triggers and bindings

In the next section, we will define more triggers in Azure Functions.

Triggers

A **trigger** defines how a function is invoked; it provokes the running of a function. Triggers can include the following:

- **Blob Storage**: When a file or folder is uploaded or changed in storage, this will invoke your function
- **HTTP**: Invokes your function similar to a REST API
- **Queue in a service bus**: When items exist in a queue, it invokes your function
- **Timer**: This will invoke your function at scheduled intervals

In the next section, we will define the bindings in Azure Functions further.

Bindings

Binding to a function is a method of declaratively connecting another resource to the function. Bindings can be connected as input bindings, output bindings, or both. Some examples of bindings include the following:

- **Cosmos DB**: Connect to the database to be able to load or save files easily

- **Table Storage**: This uses the key/value storage from your functions app

- **Queue Storage**: This retrieves one or more items from a specific queue, or places a new one in the queue

You can use different bindings by mixing or matching them according to your scenario.

Order processing scenario using Azure Functions

A healthcare company deploys optimized solutions to improve healthcare access for patients. It offers multiple applications (mobile, web, and desktop applications) for doctors and hospitals. It has a web and mobile application called **EparaMed**. This is an online drugstore that offers several services. Customers log in to purchase their products, and these products can be shipped or picked up.

The following is the workflow of an order processing scenario using Azure Functions:

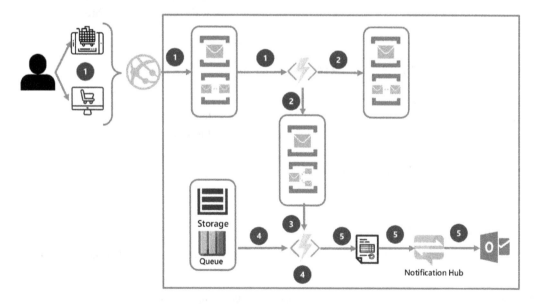

Figure 4.2 – Order processing scenario

The previous figure presents the workflow of an order processing scenario.

Let's describe the workflow. We have all these actions:

1. When a customer places a new order, the application (web or mobile) submits a message to the **service bus queue**. The message includes the order details.

2. Azure Functions separates the orders placed for the products to be submitted to the topic, and the other orders are sent to the service bus queue.

3. The product order will further be separated based on order category using **topic subscription rules**. This topic filters, by category, the orders into two subscriptions using **subscription rules**.

4. Azure Functions retrieves the price of the ordered product from the storage table and calculates the price of the order.

5. An invoice will be created based on the price, and a notification will be sent to the customer— including the product details and the invoice—using Microsoft Outlook.

This scenario is an example related to the use case of order processing, and its implementation can be found on GitHub:

```
https://github.com/PacktPublishing/A-Developer-s-Guide-to-Building-
Resilient-Cloud-Applications-with-Azure/tree/main/Chapter4
```

To accelerate application development, we can use Azure Functions as in our previous scenario. We don't need to think about the infrastructure challenges, and we have the ability to abstract the developer away from having to deal with servers—either virtual or physical—because the Azure Functions service is based on serverless architecture.

Developing Azure functions

In this section, we will discuss the key components and the structure of a function. We will see how we can create triggers and bindings and learn how we can create a function by using Visual Studio Code. We will also use Azure Functions Core Tools and connect it to the Azure service.

Azure Functions' development

A function is composed of two important elements:

- **Your code**: It can be written in any language
- **Configuration JSON file**: `function.json`, which is generated automatically from annotations in the code

The following example presents a `function.json` file. We defined the triggers, bindings, and more configuration settings needed. Every function includes only one trigger. To determine the events to monitor or the workflow of the data in the input and output of a function's execution, the runtime uses this configuration file.

The name property is used for the bound data in the function. In the following example, the name is AddAParamName:

```
"bindings":[
// Add your bindings here
{
"type": "YourBindingType",
"direction": "in",
"name": "AddAParamName",
// Add more depending on binding
}
]}
```

Let's now discuss the function app, which is a way to organize and manage your functions. When we create a function, we need an execution context in Azure where our created function will run. A function app consists of one or more individual functions that are managed, deployed, and scaled together. All functions created in a function app will share the same pricing plan, deployment method, and runtime version.

With a code editor such as **Visual Studio** (2019 or 2022) or **Visual Studio Code**, it is easy to create and test your function locally. We can even connect our local functions to live Azure services.

There are three ways to create a function app:

- In the Azure portal, write a script function
- Using the Azure CLI, create the necessary resources
- Using your favorite **integrated development environment** (IDE), such as Visual Studio 2022 or Visual Studio Code, build functions locally and publish them to Azure

In the next section, we will create an Azure Functions instance by using Visual Studio 2022. Visual Studio provides an Azure Functions template.

Creating an Azure Functions instance by using Visual Studio 2022

We will create a simple C# function that will respond to HTTP requests. Then, after testing it, we will deploy it to Azure Functions.

The prerequisites for creating the function are as follows:

- An Azure account where you are able to create any Azure resources. You can sign up for a free trial at https://azure.com/free.
- Visual Studio 2022 Community Edition, which is free to use.
- .NET Core 6, which is the target framework for the next steps.

Let's create a new project using a Visual Studio Azure Functions template.

Creating your local project

Select **Create a new project** at the bottom of the window and search using the `function` keyword in the search box, as follows:

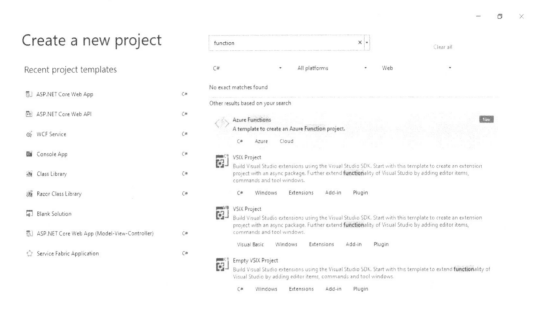

Figure 4.3 – Creating a new Functions app

Select the **Azure Functions** template and then select **New** to configure the project. This template creates a C# class library project that you can publish to a function app in Azure. We will provide the project name. Select the location and then select **Next**. You can change the solution name and check **Place solution and project in the same directory**:

Figure 4.4 – Visual Studio solution configuration

Next, we will continue configuring our project. We need to select the latest framework supported by the worker function. Azure Functions provides many templates to use depending on your scenario, as shown here:

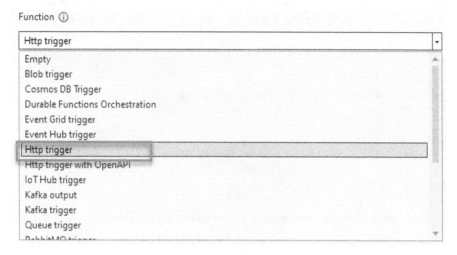

Figure 4.5 – Function templates

By default, we have **Http trigger**, but we have more, including the following:

- **Http trigger**: This is used when the code is triggered by using an HTTP request.
- **Timer trigger**: This is used to execute any batch tasks by defining a planned schedule and also for cleanup tasks.
- **Cosmos DB Trigger**: This is used to process when Azure Cosmos DB documents are added or updated in collections in the NoSQL database.
- **Blob trigger**: This is used for image resizing or to process Azure Storage blobs when they are added to containers.
- **Queue trigger**: This is used for responding to messages when they are pushed in an Azure Storage queue.
- **Event Grid trigger**: This is used to build event-based architectures if we want to respond to some events delivered to an Azure event grid.
- **Event Hub trigger**: This is used in some scenarios, such as user experience, application instrumentation or workflow processing, and **Internet of Things** (**IoT**), and to respond to events delivered to an Azure event hub.
- **Service bus queue trigger**: This is used to connect your code to other Azure services or on-premises services by listening to message queues.
- **Service bus topic trigger**: This is used to connect your code to other Azure services or on-premises services. To do that, we will subscribe to topics.

In our sample, we will use **Http trigger**. We can enable Docker.

For **Authorization level**, we can select **Function**, **Anonymous**, or **Admin**. In our case, we will select **Anonymous**, which enables anyone to call your function endpoint. In the default level—that is, **Function**—we have to present the function key in requests to access our function endpoint:

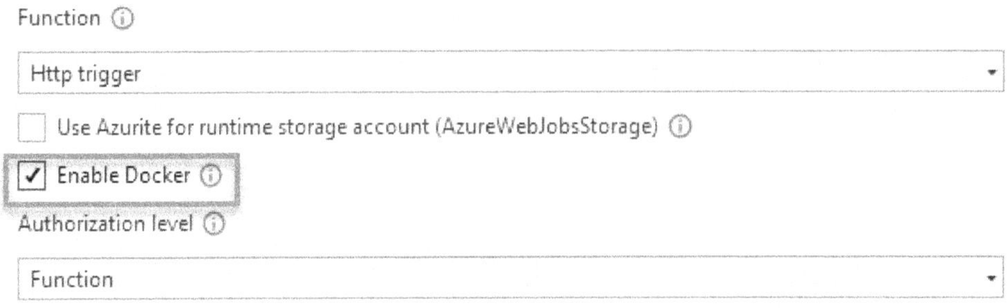

Function ⓘ

```
Http trigger                                                              ▾
```

☐ Use Azurite for runtime storage account (AzureWebJobsStorage) ⓘ

☑ Enable Docker ⓘ

Authorization level ⓘ

```
Function                                                                  ▾
```

Figure 4.6 – Enabling Docker and configuring authorization level

A function app requires a storage account that can be assigned or created when we publish the project to Azure Functions. If we select **Http trigger**, we don't need an Azure Storage account connection string. This is why we need to select **Use Azurite for runtime storage account (AzureWebJobsStorage)**.

We select the **Create** button to create the function project and **Http trigger** function:

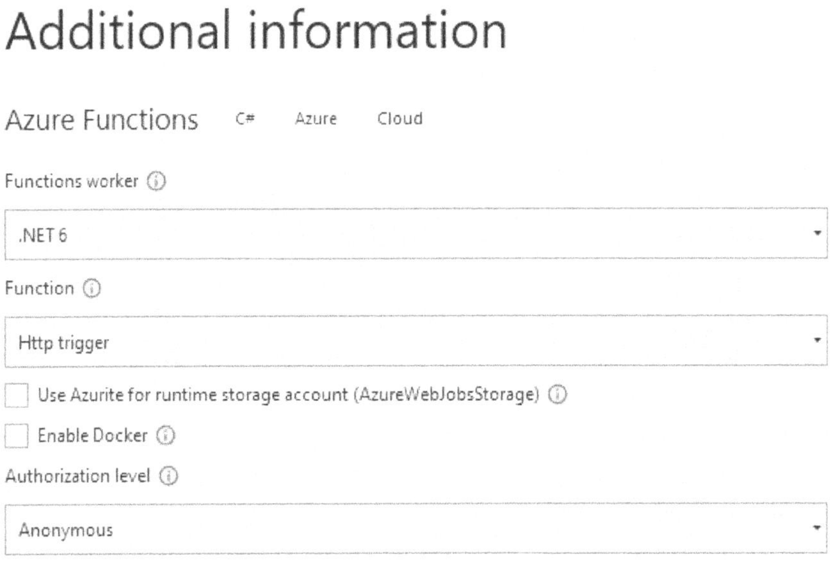

Additional information

Azure Functions   C#   Azure   Cloud

Functions worker ⓘ

```
.NET 6                                                                    ▾
```

Function ⓘ

```
Http trigger                                                              ▾
```

☐ Use Azurite for runtime storage account (AzureWebJobsStorage) ⓘ

☐ Enable Docker ⓘ

Authorization level ⓘ

```
Anonymous                                                                 ▾
```

Figure 4.7 – Configuring a functions app project

This is the structure of our functions app; we have a project that contains a class including boilerplate code for the **Http trigger** function type. An HTTP response that includes a value from the request body or query string will be sent by boilerplate code. The `HttpTrigger` attribute will specify that the function is triggered by an HTTP request. A root project folder includes all the code of the function app. We have a host configuration file named `host.json`.

The `host.json` file includes the runtime-specific configurations:

```
FunctionsAppSample
    | - bin
    | - Function1
    | | - function.json
    | - host.json
    | - local.settings.json
```

The full code sample is available on GitHub:

`https://github.com/PacktPublishing/A-Developer-s-Guide-to-Building-Resilient-Cloud-Applications-with-Azure/tree/main/Chapter4`

With the following figure, we can explore the Azure Functions structure in Visual Studio 2022:

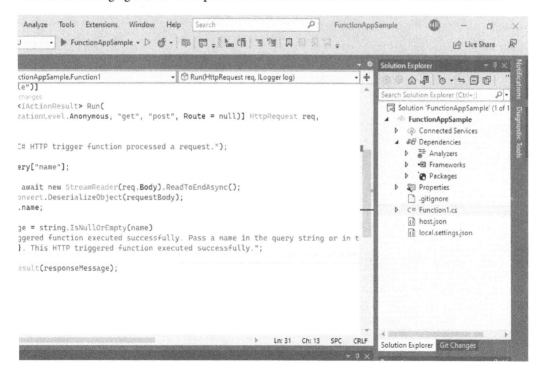

Figure 4.8 – Functions app structure

After creating the application, we run it locally. Visual Studio integrates with Azure Functions Core Tools to let you run this project on your local development computer before you publish it to Azure:

```
C:\Users\didou\AppData\Local\AzureFunctionsTools\Releases\4.10.3\cli_x64\func.exe                    —    □    ×

Azure Functions Core Tools
Core Tools Version:       4.0.3971 Commit hash: d0775d487c93ebd49e9c1166d5c3c01f3c76eaaf   (64-bit)
Function Runtime Version: 4.0.1.16815

[2022-05-25T03:48:43.529Z] Found C:\Users\didou\Documents\Book Packt\FunctionAppSample\FunctionAppSample\FunctionAppSamp
le.csproj. Using for user secrets file configuration.

Functions:

        Function1: [GET,POST] http://localhost:7071/api/Function1

For detailed output, run func with --verbose flag.
[2022-05-25T03:48:57.350Z] Host lock lease acquired by instance ID '00000000000000000000000077F04E4C'.
```

Figure 4.9 – Running the functions app locally

As we can see from the previous figure, Function1 has GET and POST methods and the URL is http://localhost:7071/api/Function1. This is the URL you can use to execute the function. So, we will copy the URL of our function from the Azure Functions runtime output, and we will paste the URL for the HTTP request into our browser's address bar. We will append the URL by adding a name for the function, as follows: localhost:7071/api/Function1?name=functions. You can see the output here:

```
Hello, functions. This HTTP triggered function executed successfully.
```

Figure 4.10 – Running the function locally

After creating and testing the Azure Functions app, we will publish it to Azure Functions.

Publishing Azure Functions to Azure

In the Visual Studio project, right-click on the project name and select **Publish**.

We will be presented with a **Publish** window; select **Azure** and click on **Next**. We will then have to select the hosting environment and the operating system (Linux or Windows). If we selected Docker, we would publish in an Azure Functions app container. In our case, we will select **Azure Function App (Windows)**:

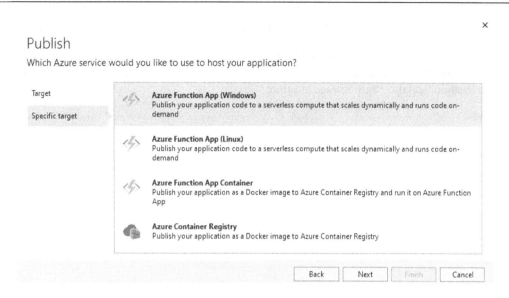

Figure 4.11 – Publishing a functions app to Azure Functions

After that, you can select an existing Azure Functions instance or create one from Visual Studio:

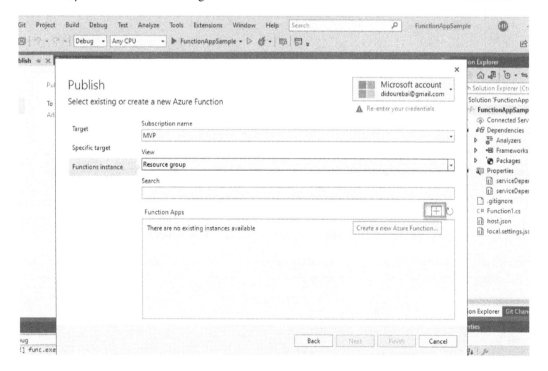

Figure 4.12 – Select an existing or create a new Azure Function window

In the previous figure, we can see the **Select existing or create a new Azure Function** window.

In the following figure, we see that to create a new function, we need to fill in all the configuration settings, such as the name and the subscription name. We will also select the resource group, the plan type, the location, and the Azure Storage location:

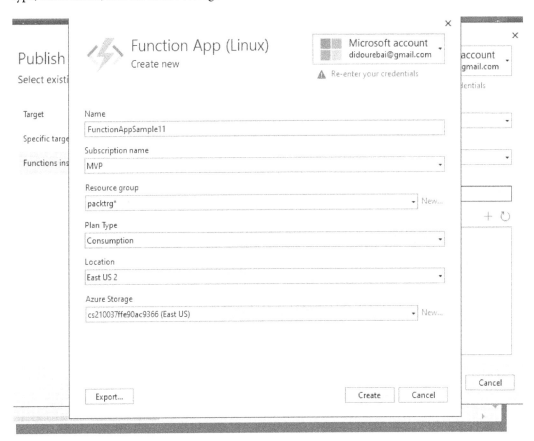

Figure 4.13 – Creating an Azure Functions instance by using Visual Studio 2022

In the previous figure, we used Visual Studio 2022 to create an Azure Function instance.

When we click on **Create**, the resource will be created in Azure and we will get the window shown in the following figure, which confirms the creation:

Publish

Select existing or create a new Azure Function

Microsoft account
didourebai@gmail.com

⚠ Re-enter your credentials

Target	**Subscription name**
	MVP ▾
Specific target	**View**
	Resource group ▾
Functions instance	**Search**
	FunctionAppSample11

Function Apps   ⊞ ↻

▲ 📁 **packtrg**
　▲ ⟨⟩ FunctionAppSample11 (Consumption)
　　▷ 📁 Deployment Slots

[Back] [Next] [Finish] [Cancel]

Figure 4.14 – Selecting an existing Azure Function instance

Then, we will click on **Publish** to publish our functions to Azure Functions. Don't forget to compile in release mode before publishing to Azure Functions.

Figure 4.15 – Publish a functions app to Azure Function App

In the **Publish** tab, we will select **Manage in Cloud Explorer**. This will open the new functions app Azure resource in Cloud Explorer.

Creating an Azure Functions instance by using Visual Studio Code

In the previous section, we created an Azure Functions instance in Visual Studio 2022 and published it.

In this section, we will create an Azure Functions instance by using Visual Studio Code.

Let's start with the prerequisites for creating this:

- An Azure account where we are able to create any Azure resources. You can sign up for a free trial at `https://azure.com/free`.

- Visual Studio Code.

- Azure Functions extensions.

- .NET Core 6, which is the target framework required for the next steps; you can download the .NET 6 SDK from the official Microsoft site: `https://dotnet.microsoft.com/en-us/download/visual-studio-sdks`.

In the next figure, we will enable Azure Functions extensions in Visual Studio Code:

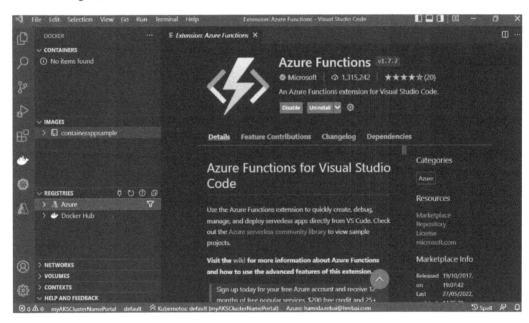

Figure 4.16 – Azure Functions extensions

We will select the **Azure** icon in the activity bar. Next, in the **AZURE** area, we will select **Create new project**:

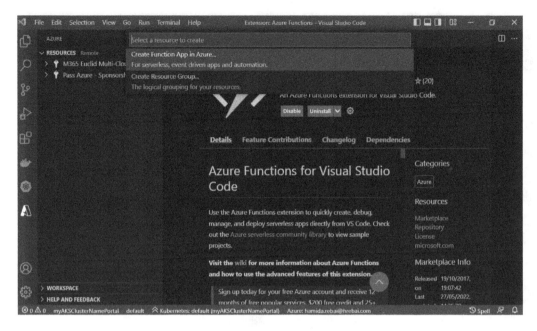

Figure 4.17 – Create Function App in Azure…

We will enter a unique name for the new function app after we select the runtime stack, which is .NET 6 in our example. After that, we will select the location and then the resource group.

We can also create a local project and publish it to Azure. To do that, we use the same **Azure** icon and select **Create new project**. After that, we need to provide all the information requested at the prompts:

- **Select a language**: Select **C#**

- **Select a .NET runtime**: Select **.NET 6**

- **Select a template for your project's first function**: Select **Http trigger**

- **Provide a function name**: Type HttpExample1

- **Provide a namespace**: Type Functions1

- **Authorization level**: Select **Anonymous**, which enables anyone to call your function endpoint

- **Select how you would like to open your project**: Choose **Add to workspace**

If we need to execute the function, we will go to the **AZURE** area, and expand **Local Project | Functions**. Right-click the function name and select **Execute Function Now…**:

Figure 4.18 – Execute function features

Previously, we created an Azure Functions app using Visual Studio 2022 and Visual Studio Code, but we can also use the Azure portal to create Azure functions and add and run multiple functions without implementing an application.

Creating an Azure Functions app in the Azure portal

We can create our Azure Functions app using different tools. In this section, we will create an Azure Functions app by using the Azure portal.

To create an Azure Functions app, we will open the Azure portal and select the resource related to Azure Functions. The following figure presents the different configuration settings:

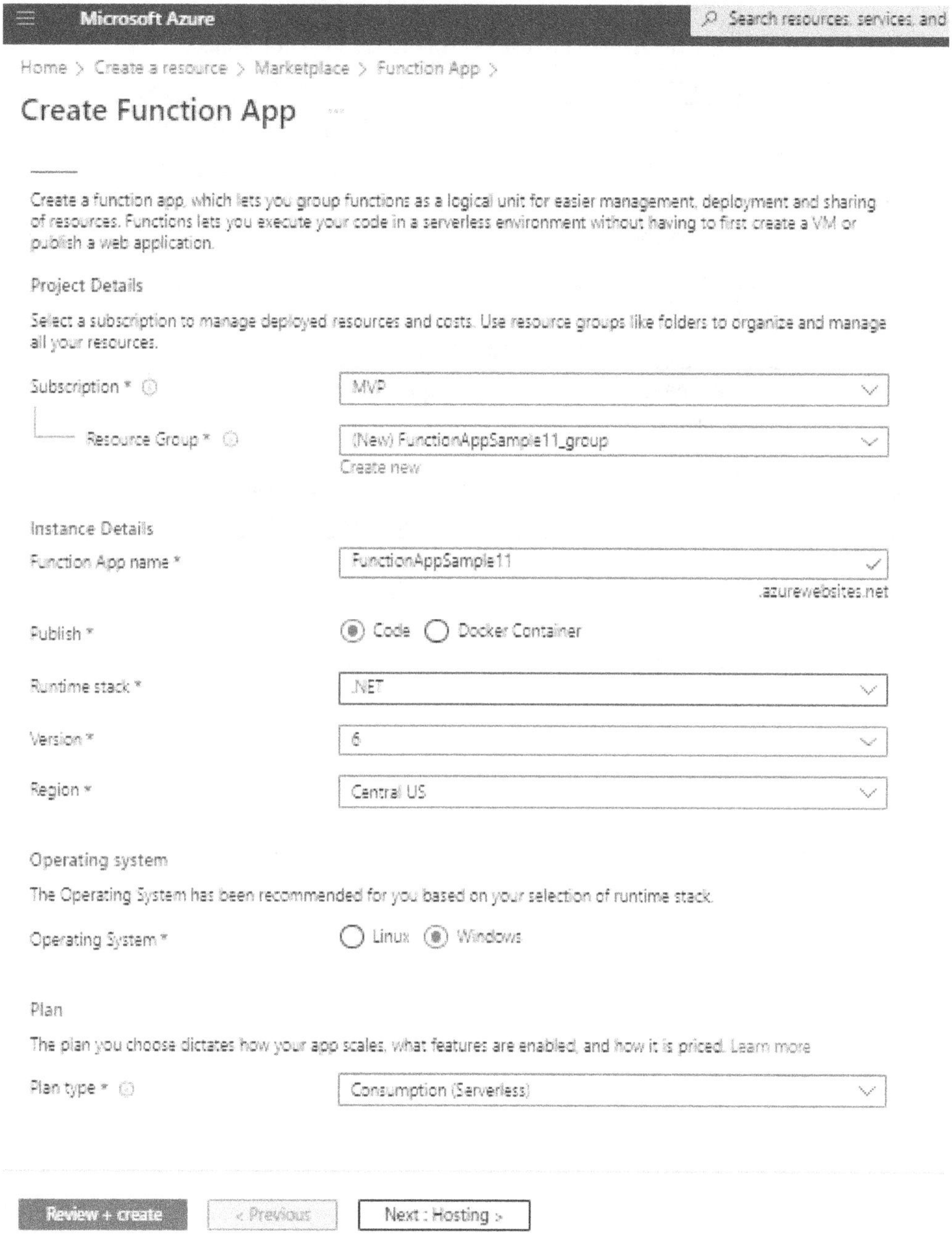

Figure 4.19 – Create Function App window

We will fill in all the information required to create a functions app in the following table (see *Figure 4.18*):

Setting	Description
Subscription	The subscription to use in order to create a new function app.
Resource group	You can create a new resource group or select an existing one.
Function app name	The name used to identify the function app. Characters that can be used are a-z (case insensitive), 0-9, and -.
Publish	Select code files or a Docker container.
Runtime stack	Select the runtime that will support the function—for example, the C# class library, the C# script, JavaScript, PowerShell, Java, and TypeScript.
Version	Installed runtime version.
Region	Select a region near you or near other services that the function will access.
Plan type	We have three plans: Consumption plan, Premium plan, and Dedicated plan. If you need more details related to these plans, you can check this link: `https://learn.microsoft.com/en-us/azure/azure-functions/functions`-scale. We will use the Consumption plan and pay only for functions app executions.

Table 4.1 – Settings and description to create an Azure Functions app

When the functions app is created, we will check the resource group. These resources are added to the resource group:

- Application Insights
- Storage account
- Function app
- App Service plan

In the following figure, we will select **Functions** under the **Functions** section:

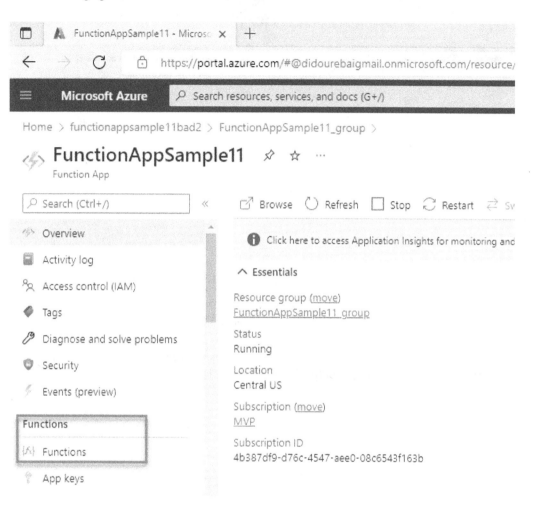

Figure 4.20 – Functions settings

We will select an existing functions app, but we can add more functions:

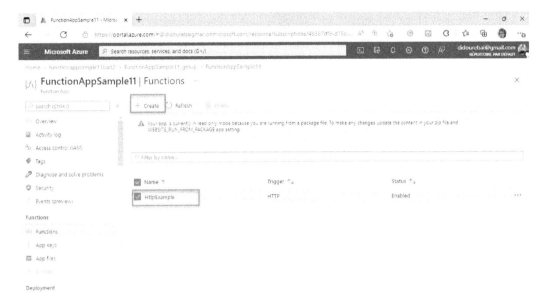

Figure 4.21 – Selecting or creating a function

We can only use the portal to configure our Azure Functions. We can create a message in a queue when a function is triggered using an output binding with an HTTP request:

Figure 4.22 – Creating a new function

Select **Code + Test**. We have `function.json`, and to run it, we have to select **Test/Run**:

Figure 4.23 – Testing and running a functions app

Select **Integration**, and then select **+ Add output**:

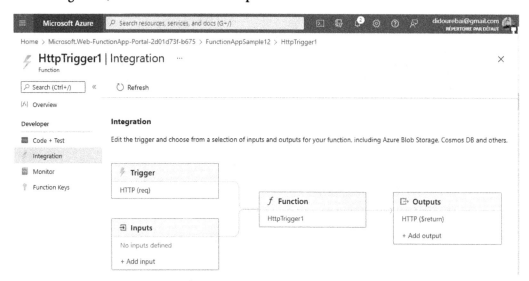

Figure 4.24 – Integration

We can add an input as a binding by selecting **Add input**. The input can be **Azure Blob Storage**, **Azure Cosmos DB**, **Azure Table Storage**, **Durable Client**, **Orchestration Client**, or **SignalR Connection Info**:

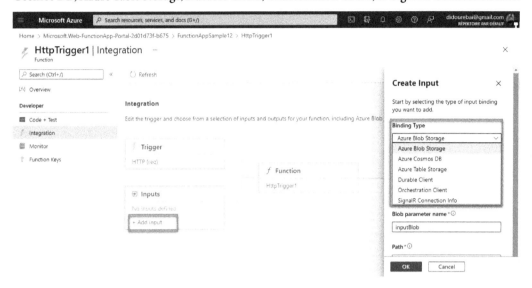

Figure 4.25 – Adding an input binding

For output binding, we can select more types: **Azure Blob Storage**, **Azure Cosmos DB**, **Azure Event Hubs**, **Azure Queue Storage**, **Azure Service Bus**, **Azure Table Storage**, **HTTP**, **Kafka Output**, **SendGrid**, **SignalR**, and **Twilio SMS**:

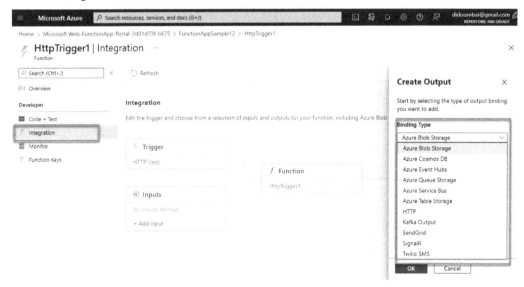

Figure 4.26 – Adding an output binding

In this section, we added input and output bindings in an Azure Functions app.

When choosing between regular functions and durable functions, one of the key factors in the decision is whether or not you're implementing a sequential, multi-step workflow as part of your application. If you have functionality in your app that depends on previous steps or specific states to trigger the next steps, durable functions make several tricky parts of these workflows much easier. So, what are durable functions, and how can we develop them?

Developing durable functions

In this section, we will describe the diverse types and patterns used for durable functions, and we will explain the use of durable orchestrations, timers, and events. Azure Durable Functions is an extension of Azure Functions that is used to write stateful functions. Durable Functions supports multiple languages, such as C#, JavaScript, Python, F#, and PowerShell.

Later in this chapter, we will describe the common application patterns that can benefit from durable functions:

- Function chaining
- Fan-out, fan-in
- Asynchronous HTTP APIs
- Monitor
- Human interaction

Now that we have defined durable functions and introduced the different patterns used, we will describe durable functions in more detail and the different scenarios that require their use.

Introduction to durable functions

To implement more complex scenarios with serverless functions, we need a mechanism to allow us to save the state during the execution of these functions.

This extension of functions allows us to write the stateful functions and define orchestrations of other functions. It automatically handles the state, checkpoints, and restarts. There are several benefits of using Azure Durable Functions in our projects.

We will start by building a stateful workflow to overcome the platform's inherently stateless nature. Next, the defined functions as orchestrators can invoke other functions synchronously or asynchronously. The result of the functions invoked by the orchestrator can be saved as local variables. All the progress made by the orchestrator and the rest of the functions is stored locally. We must ensure that we never lose the state. This is part of the fundamental functionality of the platform. When the activity functions are in bold, you can specify an automatic retry policy. The functions that are considered orchestrators can be **Service Orchestration**, **Resource Orchestration**, and **Lifecycle Orchestration**. There is a relationship of dependency and continuity between the functions. You can also run multiple instances of an orchestrator function in parallel to each other functions. Orchestrator functions can have multiple statuses, such as **Started**, **Consulted**, or **Canceled**. Finally, the Azure Durable Functions extension is supported by the local Azure function and the SDK development tools that you already have on your computer.

Implementing an orchestrator function

So, here, in the `FunctionAppSample` project, we will right-click and select the **Add | New Azure Function…** option and name it `FunctionOrchestrator`:

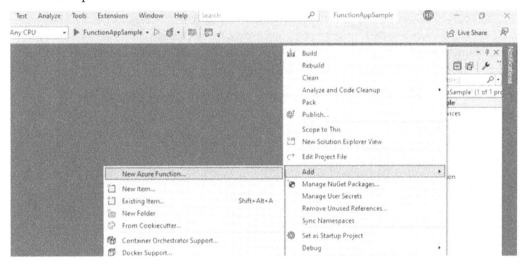

Figure 4.27 – Adding a new Azure function

Now, we will use the **Durable Functions Orchestration** template, as presented in the following screenshot:

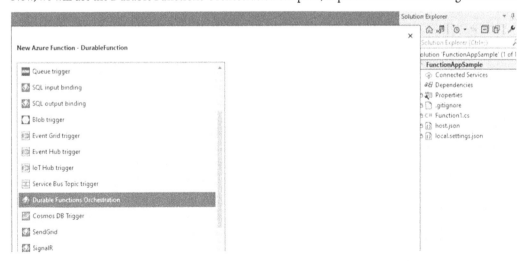

Figure 4.28 – Adding the Durable Functions Orchestration template

Don't forget to add the `Microsoft.Azure.WebJobs.Extensions.DurableTask` NuGet package, as presented in the following screenshot:

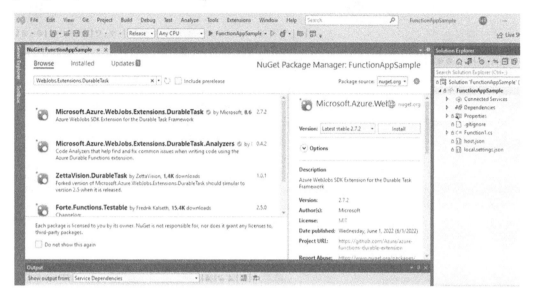

Figure 4.29 – Adding the Microsoft.Azure.WebJobs.Extensions.DurableTask package

You can implement your orchestrator according to your needs and add or delete the generated methods.

We can create an orchestrator function using the Azure portal. In Azure Functions, we will select **Functions** from the left pane, then select **Add** from the top menu.

In the search field of our new function page, we can use the `durable` keyword and then select the **Durable Functions HTTP** starter template.

We are requested to provide a function name. After, we can click on the **Create** function.

We implemented an orchestrator function using Visual Studio 2022, but let's review some of the orchestration patterns that Durable Functions provides. We will start with the function chaining pattern.

Function chaining

Function chaining is where we need to execute a sequence of activity functions following a specified order. The orchestrator function will keep track of where we are in the sequence:

Figure 4.30 – Function chaining pattern

Durable Functions makes it easy to implement this pattern in your code. Another pattern will be presented in the next section: fan-out, fan-in.

Fan-out, fan-in

Fan-out, fan-in is a pattern that runs multiple activity functions in parallel and then waits for all the activities to finish. The results of the parallel executions can be aggregated and used to compute a final result. This pattern is very hard to use without a framework such as Durable Functions:

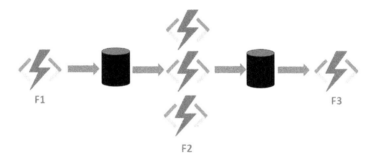

Figure 4.31 – Fan-out, fan-in pattern

This pattern is much more challenging, but the Durable Functions extension handles it with simple code.

Asynchronous HTTP APIs

Asynchronous HTTP APIs are useful when you have an API that you need to repeatedly pull for progress. We start a long-running operation, and then the orchestrator function can manage the pulling until the operation has completed or timed out:

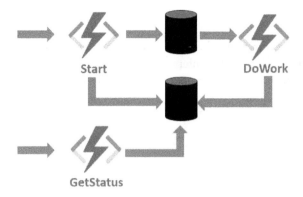

Figure 4.32 – Asynchronous HTTP APIs pattern

The Durable Functions runtime manages the state. We are not going to implement the status-tracking mechanism. In the next section, we will explore the monitor pattern.

Monitor pattern

The monitor pattern implements a recurring process in a workflow, looking for a change in state. We use this pattern to repeatedly call an activity function that's checking to see whether certain conditions are met:

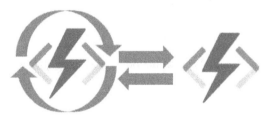

Figure 4.33 – Monitor pattern

To create multiple monitors in order to observe arbitrary endpoints, we use Durable Functions with a few lines of code. We will see the last pattern, human interaction, in the next section.

Human interaction

Human interaction is very common in business workflows to require manual approval where the workflow must pause until a human has interacted in some way. This pattern allows workflows to wait for certain events to be raised and then optionally perform a mitigating action if there's a timeout waiting for the human interaction:

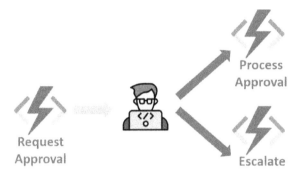

Figure 4.34 – Human interaction pattern

A durable timer is used by the orchestrator function to request approval. If a timeout occurs, the orchestrator will escalate. But the orchestrator keeps waiting for any external event—for example, a notification generated by human interaction.

Summary

During this chapter, we learned about Azure Functions. We developed Azure Functions and durable functions and learned how we deploy them in Azure Functions. We explored how we create a function using the portal, Visual Studio 2022, and Visual Studio Code.

In the next chapter, we will learn about Azure Service Fabric's main benefits, as well as how to build Service Fabric applications for the cloud or on-premises.

Questions

1.   Do we need to add more packages for Durable Functions?

2.   What are durable functions in Azure?

Develop an Azure Service Fabric Distributed Application

Enterprises face challenges in moving to the cloud and navigating cloud-based application optimization by considering latency, monitoring, and governance issues.

When we build a complex, scalable distributed application and would like to use a solid platform to host, build, and run it with high availability and low latency, we can use Azure Service Fabric because it is suitable for applications looking for robustness, reliability, scalability, and resilience.

In this chapter, you will learn about the essential concepts of Azure Service Fabric, the main benefits, as well as how to build Service Fabric applications for the cloud or on-premises. We will deploy our solution on Azure Service Fabric.

In this chapter, we're going to cover the following main topics:

- Exploring Azure Service Fabric
- The Azure Service Fabric development environment
- Exercise 1 – creating a Service Fabric cluster using the Azure portal
- Exercise 2 – creating a Service Fabric cluster using the Azure CLI
- Exercise 3 – scaling an Azure Service Fabric cluster
- Exercise 4 – creating a .NET Service Fabric application
- Exercise 5 – deploying an app to a Service Fabric managed cluster and containers

Exploring Azure Service Fabric

We can consider three categories of Azure services to deploy cloud-based applications:

- **Platform as a Service (PaaS)**: This provides a hosted environment in which users can meet various computing needs. Users are able to develop, manage, and run applications on the platform while the PaaS provider manages the infrastructure. Azure offers PaaS services such as **Service Fabric** and **Azure App Service**.

- **Container as a Service (CaaS)**: The emergence of virtualization has transformed the IT industry, allowing people to run different operating systems simultaneously on a single computer. This has improved efficiency and improved performance. However, virtualization is not without its drawbacks. A separate operating system is required for a virtual environment to function. As a result, it takes up a lot of disk space. CaaS was born as a means of offloading hardware resources. CaaS providers give users access to containerized applications or clusters. Azure offers CaaS services such as **Azure Service Fabric**, **Azure Container Instances**, **Azure Kubernetes Service**, which is a container orchestrator containing microservices that are easy to manage, and **Azure Container Apps**.

- **Function as a Service (FaaS)**: FaaS is a type of cloud computing service that allows code to be implemented in response to events without extensive code infrastructure changes. It is recommended for users who only need the cloud for individual functions within the app. Users do not need to build the infrastructure normally required for app development.

The following diagram presents the different services used for every type: PaaS, CaaS, and FaaS. Azure Service Fabric can be used as a PaaS and for containerized applications as a CaaS.

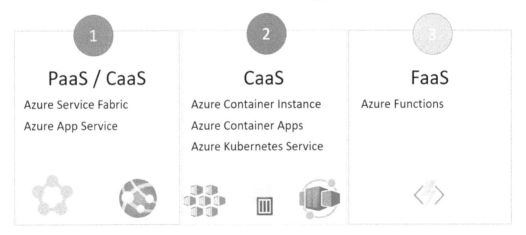

Figure 5.1 – Azure container services

In this section, we discussed the different Azure services to build and deploy container applications. In the next section, we will define Service Fabric and the different services using Service Fabric.

Definition

Azure Service Fabric allows you to create modern, resilient, and scalable software solutions, with features required to build a robust system, such as autoscaling, health monitoring, and fault tolerance.

The Service Fabric platform offers a programming model that you can use to create reliable services or components based on the actor model. For containerized applications, Service Fabric also supports Docker containers so that you can use the platform as an orchestrator. Azure Service Fabric is able to run on any operating system (Windows or Linux). Additionally, it can run in any environment, on-premises as a private cloud or on a public cloud such as Azure or any other cloud provider, such as AWS or Google Cloud Platform, or in the local development environment of a developer. It offers a complete set of development tools and **software development kits (SDKs)** to start building solutions on the platform very quickly. These tools can be used in a development environment or in a production environment. The cloud has allowed the evolution of software architectures and development platforms for distributed systems. Azure Service Fabric is not new. It is an open source project that is used in many services in the Azure infrastructure as well as other Microsoft services. In fact, it has existed since the beginning of Azure, when it was still called Windows Azure. Within Microsoft, this platform is strategic and fundamental since many cloud services use it, such as Azure SQL Database, Azure Cosmos DB, Azure DevOps, Dynamics 365, Power BI, and many more. The following diagram depicts the different Microsoft services and tools that use Service Fabric.

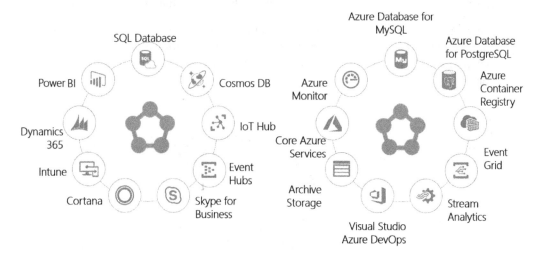

Figure 5.2 – Powering Azure and Microsoft services with Service Fabric

But how does Azure Service Fabric work? In the next section, we will present the core functionality of Azure Service Fabric, which is based on two essential concepts: clusters and nodes.

Clusters and nodes

A Service Fabric cluster is a network-connected set of **virtual machines** (**VMs**) in which your microservices are deployed and managed.

A cluster is a set of connected nodes that are VMs, which form a highly available and highly reliable execution environment in order to run services and applications.

Service Fabric supports any number of nodes in a cluster but the number depends on the use of the cluster. For example, in the development phase, one node is used, but in production, you need to have more than five. To deploy a Service Fabric instance in a cluster, we have to install it in one of the nodes. Then, the Service Fabric instance will replicate itself to the remaining nodes in the cluster automatically.

Every node has an operating system, Windows or Linux, with a runtime. A node can host multiple applications – it depends on the resource available in the VM.

In an Azure Service Fabric deployment, the load balancer is used to detect healthy nodes by sending a request to them and avoids any unhealthy nodes. To deploy an application, we can use the load balancer because it routes the request to one of the available nodes. Then, Service Fabric will replicate the application to the remaining nodes. Service Fabric acts like a scheduler. In the end, once the application is deployed, the load balancer will route the requests to one of the healthy available nodes so that the application can respond. This is why you can run Service Fabric applications in any operating system.

The following figure shows a stateless service with four instances distributed across a cluster using one partition. Note that partitioning is a concept that enables data storage on a local machine. It consists of decomposing state, which is data, and also computer resources into smaller units to ensure scalability and performance.

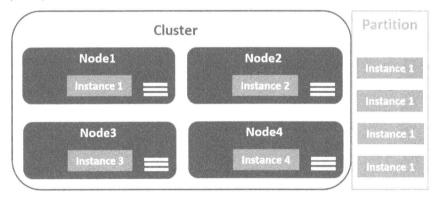

Figure 5.3 – An instance of a stateless service across a cluster

We have three main features that emerge from this scenario, thanks to redundant servers: we ensure availability, the creation of several replicas of data, as well as an evolution in terms of the partitioning of services ensuring reliability. Partitioning is a key feature of the reliable services programming model. Service Fabric is undoubtedly an excellent option for hosting and running the most critical software solutions.

The differences between Service Fabric and Kubernetes

Containers are used in cloud-native applications. This technology has impacted the software development industry and has changed how systems are designed, operated, distributed, and monitored. If we have multiple containers to manage, we need an orchestrator so that we'll have various container orchestrators. One of the main features of an orchestrator is container life cycle management, such as container provisioning, the scaling of existing containers, and resource allocation. Kubernetes is one of the most widely used container orchestrators in the industry.

The main difference between Kubernetes and Service Fabric is that Service Fabric has three mechanisms to deploy and run applications:

- **A propriety programming model that you can optionally use in your applications**: This programming model allows you to build a stateful or a stateless reliable service, and also reliable actors. The actor model is based on having distributed objects that communicate with each other through messages.

- **The use of containers**: Both Linux and Windows containers are supported.

- **The use of any secure build with any platform and any language, with no modification required at all**: Kubernetes supports both Linux and Windows containers. Likewise, Microsoft Azure includes Azure Kubernetes Service, which is a fully managed platform for Kubernetes.

The Azure Service Fabric development environment

We have many different options to adopt Service Fabric. Service Fabric is a free, open source project whose source code is available on GitHub. Service Fabric supports three different execution environments:

- **On-premises environment**: You can install Service Fabric locally on your servers or VMs. This scenario is suitable for the first cloud migration of your workloads.

- **Public cloud**: Azure Service Fabric allows cluster creation. Note that we have the same Service Fabric functionalities as the previous environment because we can use the same runtime and SDK.

- **Public cloud with other providers**: We can install Service Fabric on any VM using any operating system.

We can consider another environment using containers as if they were nodes of a cluster. As we said before, Service Fabric is open source so it is free. However, if you deploy to a cloud provider, then you pay for the computing resources that you provision.

Let's now talk about the development environment and the prerequisites. We will present the different tools used to build and deploy a .NET application. The Service Fabric SDK also supports the Java programming language and is able to host any executable, regardless of the platform or language used to build it.

We must meet these prerequisites in order to be able to configure the development environment to create a Service Fabric cluster:

- Windows 10 or 11, which supports Hyper-V, is required for Docker installation
- Visual Studio 2022 or 2019, or Visual Studio Code
- The .NET Core SDK
- Microsoft Service Fabric SDK
- Docker tools for Windows
- The Azure CLI

You can download Visual Studio from `https://visualstudio.microsoft.com/`. The Community Edition is always free to use and includes Service Fabric templates. You can also use the other editions.

To download and install the Service Fabric SDK, use this link: `https://www.microsoft.com/web/handlers/webpi.ashx?command=getinstallerredirect&appid=MicrosoftAzure-ServiceFabric-CoreSDK`.

You can download Docker for Windows from the Docker website.

Use this link to install Docker Desktop and to create an account in Docker Hub: `https://www.docker.com/get-started/`.

Finally, we will install the Azure CLI tool; you can use this link to install it for Windows: `https://aka.ms/installazurecliwindows`.

Now, we have all the required software to build a Service Fabric application.

Let's check the installation now. We will press **Start** and type `fabric` to search for Service Fabric Local Cluster Manager. Then, press *Enter* in order to set up a local cluster as presented in the following figure:

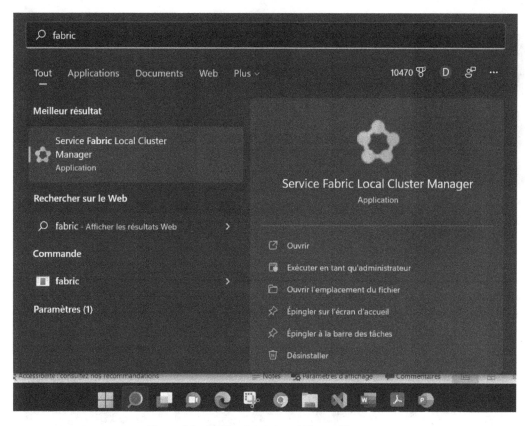

Figure 5.4 – Service Fabric Local Cluster Manager

From the system tray, we will right-click the Service Fabric icon, as presented in the next screenshot. Then, we will navigate to **Setup Local Cluster** to select the number of nodes: we have the **1 Node** or **5 Node** options for a Windows or Linux environment. When we select the number of nodes to use, we wait for a few seconds until we see the **Service Fabric Local Cluster Manager setup completed successfully** notification.

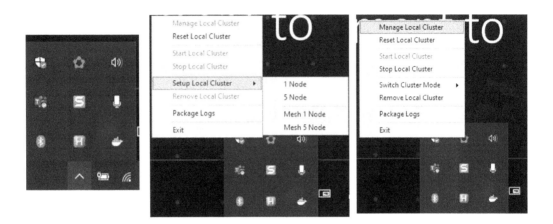

Figure 5.5 – Setup Local Cluster

We will select **Manage Local Cluster** to open **Service Fabric Explorer**. We can take a look at the cluster configuration. I selected a single node, for testing purposes, with less memory. In the next figure, we can see that our local environment uses **1 Node** and is just a virtual node running on my machine.

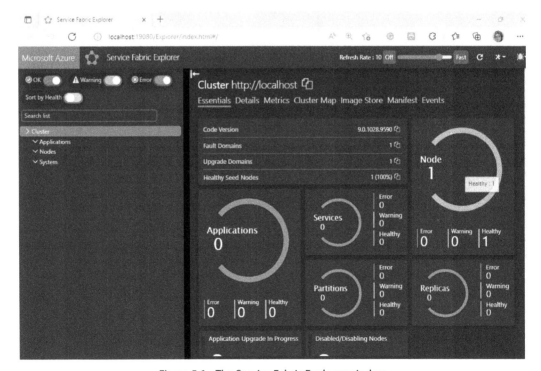

Figure 5.6 – The Service Fabric Explorer window

You can create an application using the Service Fabric template in Visual Studio 2022. Select any application type, then compile it, and the application will be added to the explorer. It's another method to verify whether your cluster is configured or not.

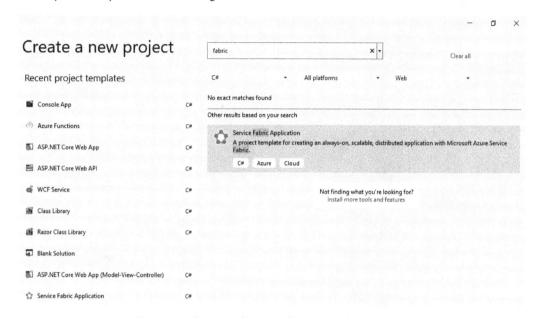

Figure 5.7 – Service Fabric template in Visual Studio 2022

We can deploy any application locally and we can debug the clusters using Service Fabric Explorer. Service Fabric Explorer is a monitoring tool used by developers. We can also deploy our cluster remotely on Azure Service Fabric. We will see in the next section how we can create an Azure Service Fabric cluster.

Exercise 1 – creating a Service Fabric cluster using the Azure portal

In Azure, we will navigate to the main page of the Azure portal. From there, we will select the **Create a resource** button, and then we will search for `Fabric`. Or, just select **Containers** on the left of the window and, in the end, we will find **Service Fabric Cluster**, as shown in the following figure. Let's select **Service Fabric Cluster**, then click on the **Create** button.

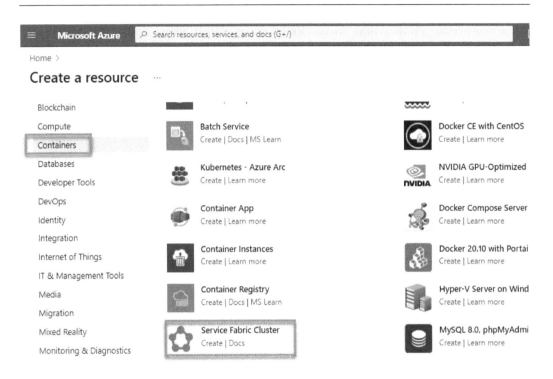

Figure 5.8 – Select Service Fabric Cluster

Next, we will specify some general information about the cluster in the **Basics** blade. We will select the subscription and an existing resource group; you can also add a new one. We're going to use `hospitalmanagement` as the name. Remember to select a location close to your current location. Also, you can specify the operating system that you want to use. In this case, we're going to use **WindowsServer2019-Datacenter-with-Containers**. However, you can select any other operating system version that you want, such as Linux. We will set the credentials for the VMs.

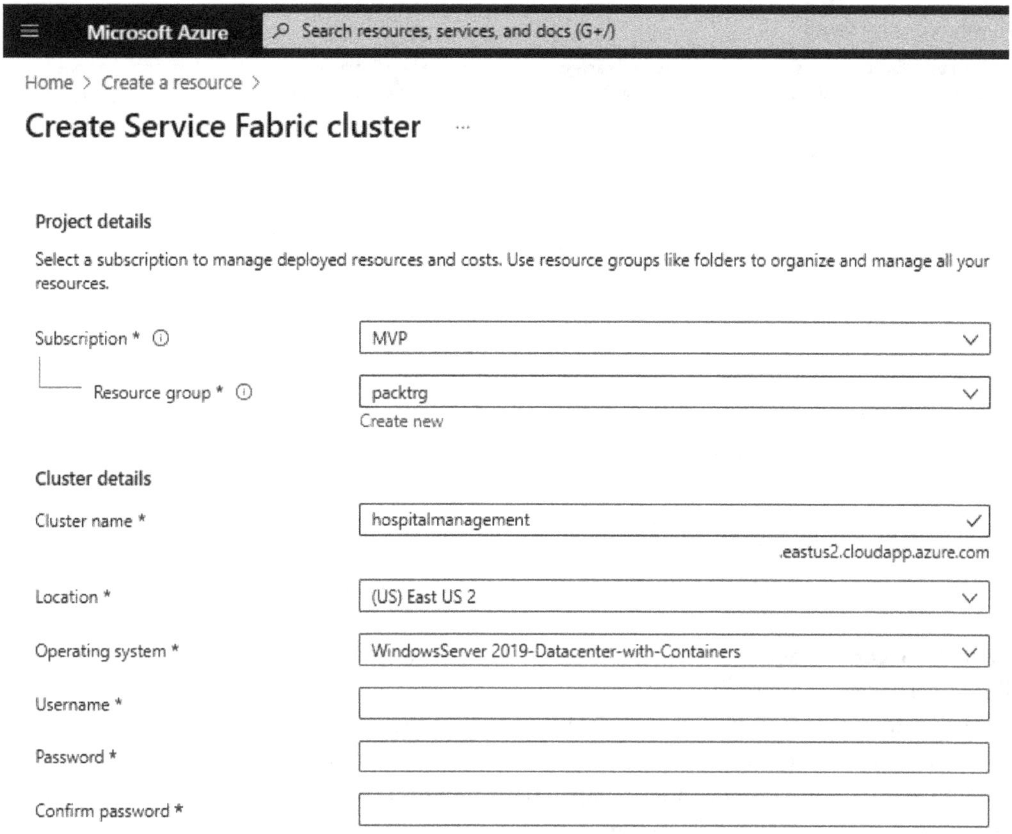

Figure 5.9 – Step 1 – Create Service Fabric cluster

Now, we will complete the cluster configuration related to the primary node. We will set up the initial VM scale set capacity, then we will define the node types. **Node types** allows you to define different hardware configurations and also allows you to specify a different scale. In our case, we will select the default values.

Primary node type details

Node types define the scale sets that will be used to manage your cluster. A Service Fabric cluster can consist of more than one node type. In that event, the cluster consists of one primary node type and one or more non-primary node types. Configure your primary node type here. Learn more about node types

Initial VM scale set capacity *

```
●●●○●●●●●●●●●●●●●●●●●●●●●●●●●●●●●●●●●●●●●●●●●●●●●●●●
```
| 5 |

50

Node types * ○

Standard D2s v3
Select VM size

Security

Service fabric requires a primary certificate to authenticate the nodes in a cluster and provide secure communication. This certificate must contain a private key and have a subject name that matches the domain used to access the cluster. You can create a new certificate now or select an existing one from an Azure Key Vault. The certificate must be stored in an Azure Key Vault in the same subscription and region as the cluster. Learn more about Service Fabric security

Key vault and primary certificate *

Select a certificate

Figure 5.10 – Step 2 – Create Service Fabric cluster

Now, in the **Security** blade, we're required to select or create an Azure Key Vault resource where we will store certificates. So, let's click on this option and we will create a new key vault. Let's name this `servicefabrickeycert`, for example. We will select the same resource group as we selected for Service Fabric. After, we will click on the **Create** button. So, now the Azure portal will deploy this Azure Key Vault resource because it's required in order to create the certificate and store it there.

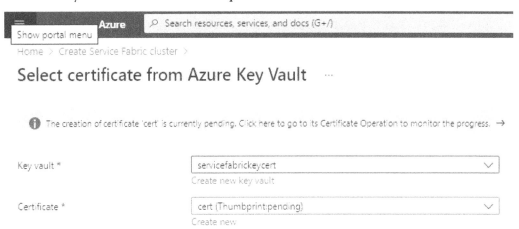

Figure 5.11 – Select certificate from Azure Key Vault

After the key vault is provisioned, you will see a message that says that it's not enabled for deployment, so we need to click on the **Edit access policies** button. The key vault has to be enabled for deployment.

To enable the key vault, we need to edit the access policies for the key vault by selecting the **Enable access to Azure Virtual Machines for deployment** option, located in the **advanced access policies** section, and clicking **Save**.

Key vault and certificate

Keyvault: servicefabrickeycert1
Primary certificate: servicefabriccert
Select a certificate

❌ The key vault you have selected is not enabled for deployment. To enable it, you will need to edit the access policies for the key vault by selecting the option for 'Enable access to Azure Virtual Machines for deployment' located in the 'advanced access policies' section and clicking 'Save'. if you do not have permissions to edit the selected key vault, you will need to contact the owner of the key vault to make the change or choose a different key vault.

Edit access policies for servicefabrickeycert1

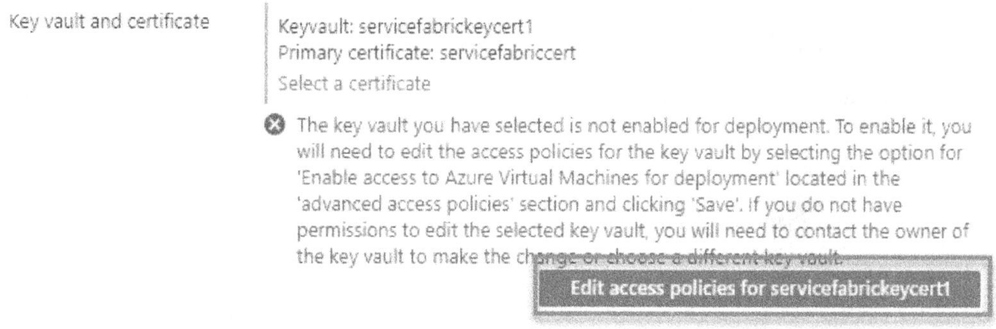

Figure 5.12 – The edit access policies for the key vault and the certificate

> **Important note**
>
> The certificate must be stored in Azure Key Vault in the same subscription and region as the cluster.

In the **Node types** tab, we can specify how many VMs we want in our scale set. As you can see, here we're using the value of five. Also, you can specify a single-node cluster.

Create Service Fabric cluster ···

| Basics | Node types | Security | Advanced | Tags | Review + create |

Node types define the scale sets that will be used to manage your cluster. Learn more about node types ↗

Single node cluster ⓘ   ○ Yes ◉ No

╋ Add   🗑 Delete

Node type name	Initial VM scale set capacity	Virtual machine size
☐ Type908 (primary)	5	Standard_D2s_v3

Figure 5.13 – Step 3 – Create Service Fabric cluster – Node types tab

This process will take a few minutes to complete. If we want to simplify the creation of a Service Fabric cluster using interactive mode, we will use the portal. However, we can also create it using the Azure CLI. This is what we will look at in the next section.

Exercise 2 – creating a Service Fabric cluster using the Azure CLI

Another option that we can use to create Azure Service Fabric clusters is the Azure CLI tool. You can open PowerShell if it is installed on your machine, or open Azure Cloud Shell using this link: `https://shell.azure.com/`.

You can find the entire sample command line on GitHub by following this link: `https://github.com/PacktPublishing/A-Developer-s-Guide-to-Building-Resilient-Cloud-Applications-with-Azure/tree/main/Chapter5`.

We will use PowerShell; we will type this command line to create a new cluster:

```
az sf cluster create -g packtrg -c hospitalmanagement -l estus
--cluster-size 5 --vm-password Password#1234 --certificate-
output-folder MyCertificates --certificate-subject-name
hospitalmanagement
```

We can use **Key Vault Certificate** and a custom template to deploy a cluster as follows:

```
az sf cluster create -g packtrg  -c hospitalmanagement -l
estus--template-file template.json \
    --parameter-file parameter.json --secret-identifier
https://{KeyVault}.vault.azure.net:443/secrets/{MyCertificate}
```

To create a new Azure Service Fabric cluster, we can use one or more properties mentioned here:

```
az sf cluster create --resource-group
                      [--cert-out-folder]
                      [--cert-subject-name]
                      [--certificate-file]
                      [--certificate-password]
                      [--cluster-name]
                      [--cluster-size]
                      [--location]
                      [--os {UbuntuServer1604,
WindowsServer1709, WindowsServer1709withContainers,
WindowsServer1803withContainers,
WindowsServer1809withContainers,
WindowsServer2012R2Datacenter, WindowsServer2016Datacenter,
```

```
WindowsServer2016DatacenterwithContainers,
WindowsServer2019Datacenter,
WindowsServer2019DatacenterwithContainers}]
                    [--parameter-file]
                    [--secret-identifier]
                    [--template-file]
                    [--vault-name]
                    [--vault-rg]
                    [--vm-password]
                    [--vm-sku]
                    [--vm-user-name]
```

Once created, you can list the cluster resources using this command or just check the Azure portal:

```
az sf cluster list [--resource-group]
```

After a cluster's creation, sometimes we need to scale it. In the next section, we will learn how to scale an Azure Service Fabric cluster.

Exercise 3 – scaling an Azure Service Fabric cluster

There are three mechanisms to scale a Service Fabric cluster.

Manual scaling

The first one is to manually specify the number of node instances that you want. So, let's navigate to the VM scale set resource. We need to select the resource group that includes the Service Fabric instance. Then, select **Virtual machine scale set** as the type, as presented in the following screenshot.

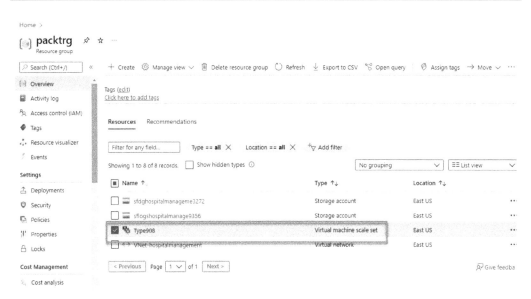

Figure 5.14 – Virtual machine scale set resource type

We will click to open it. In the **Settings** section, you will find the **Scaling** option.

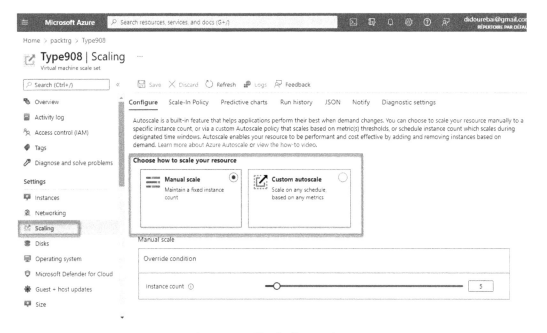

Figure 5.15 – The Scaling settings

As shown in the preceding screenshot, the default option is **Manual scale** but you can specify the number of node instances. In the figure, we have **5** node instances, but we can set it to **20**, for example. After saving this option, the VM scale set will increase from **5** to **20** instances.

Custom autoscaling

The second option is to use custom autoscaling. If you select this option, you can specify a rule for scaling in the **Custom autoscale** window, as shown in the following screenshot:

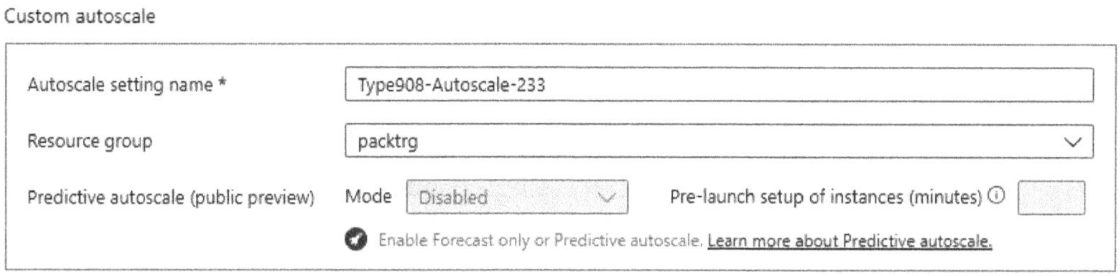

Figure 5.16 – The Custom autoscale window

We can add a scale condition. We have an auto-created scale condition by default, but we can combine multiple conditions to automate the scaling:

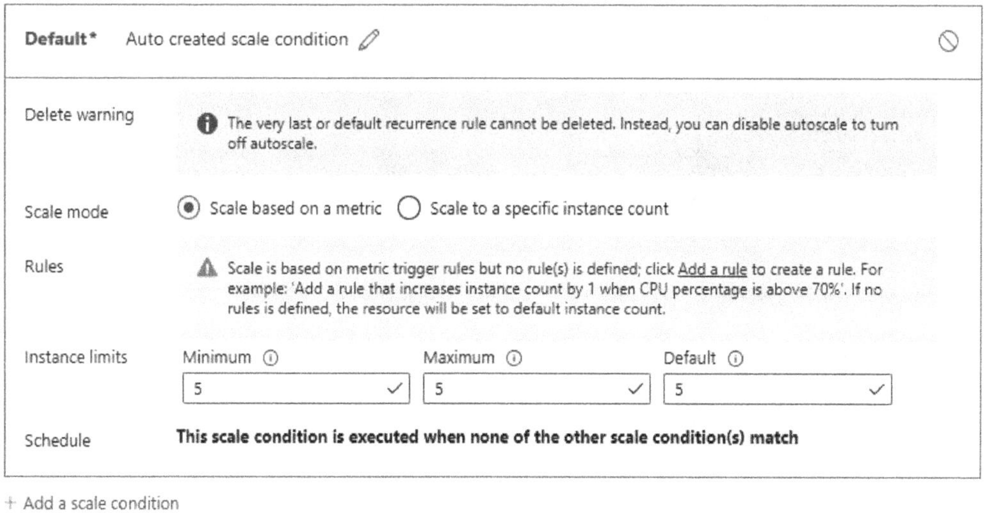

Figure 5.17 – Create a scale condition

The rules are based on a metric that you choose to determine whether the cluster needs to scale up or down. For example, based on the percentage CPU value in the last 5 minutes, if it's greater than 75, we will increase the count by *1*. You can save at the end by clicking on the **Save** button.

Coding your scaling

The third option is to scale programmatically. In other words, you create code that increases or decreases the number of instances in your cluster.

Having created our cluster and configured the scaling, we will now learn how we can create a .NET Service Fabric application.

Exercise 4 – creating a .NET Service Fabric application

Service Fabric supports multiple application models. It's a flexible technology that allows you to host and execute any workload. After installing the Service Fabric SDK, you will be able to use its available templates.

When we run Visual Studio 2022 and select **Service Fabric template**, many models appear, as shown in the screenshot:

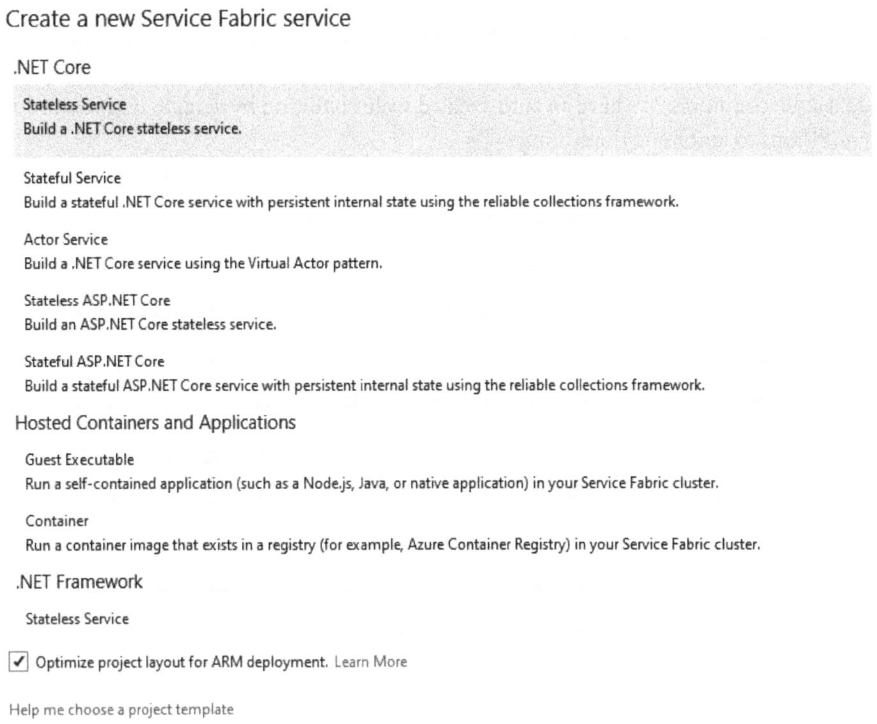

Create a new Service Fabric service

.NET Core

Stateless Service
Build a .NET Core stateless service.

Stateful Service
Build a stateful .NET Core service with persistent internal state using the reliable collections framework.

Actor Service
Build a .NET Core service using the Virtual Actor pattern.

Stateless ASP.NET Core
Build an ASP.NET Core stateless service.

Stateful ASP.NET Core
Build a stateful ASP.NET Core service with persistent internal state using the reliable collections framework.

Hosted Containers and Applications

Guest Executable
Run a self-contained application (such as a Node.js, Java, or native application) in your Service Fabric cluster.

Container
Run a container image that exists in a registry (for example, Azure Container Registry) in your Service Fabric cluster.

.NET Framework

Stateless Service

☑ Optimize project layout for ARM deployment. Learn More

Help me choose a project template

Figure 5.18 – Service Fabric template models

We will start with an application model that is considered to have reliable services. These services use the native API of Service Fabric. We have three different types of reliable services: stateless, stateful, and actors. The difference between stateful and stateless services is that in stateless services, no state is maintained in the service, but for stateful services, the state is stored with the service.

Moreover, we can use both .NET Core and .NET Framework runtimes to build reliable services. This is why we can use Service Fabric to migrate old applications. An actor is a dependent computing unit that runs on a single thread. The actor model is intended for concurrent or distributed systems since a large number of these actors can run simultaneously and independently of each other, communicating through messages.

Another application model is containers. Both Linux and Windows containers are supported in Service Fabric.

There is also the guest executable model. It allows us to bring any type of executable that we have built on any development platform and with any programming language to Service Fabric.

With these application models, we can use Service Fabric in any scenario that requires high availability, reliability, and a robust platform to deploy and run our applications.

We're going to create a sample solution based on Service Fabric's **Stateless Reliable Services**. The application is used for managing hospitals around some countries in Africa and Europe.

Creating a Service Fabric application

Let's go back to Visual Studio 2022. When we select **Create a new project**, we need to configure the project by specifying the name, the location, the solution name, and the framework. After, we will click on the **Create** button.

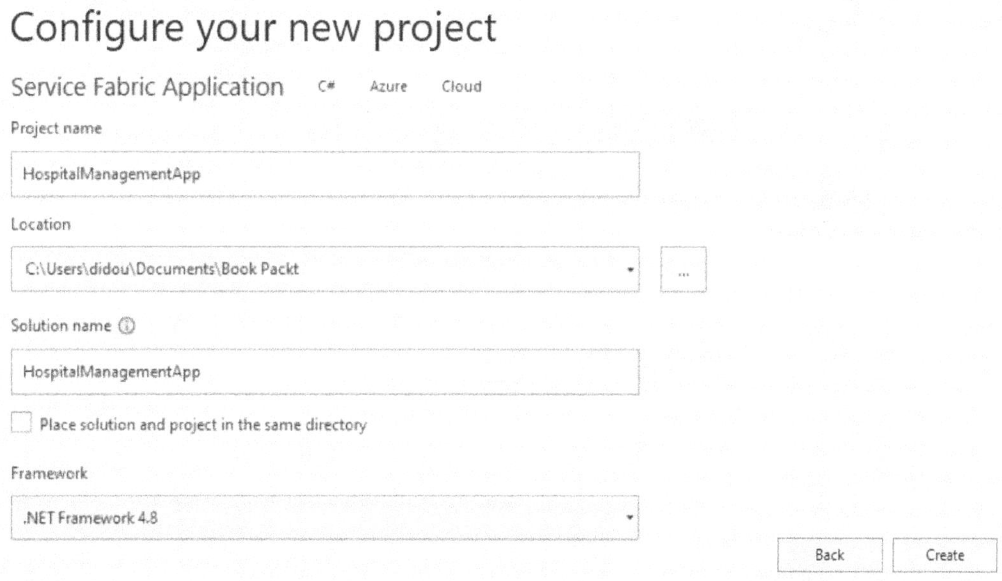

Figure 5.19 – Creating a Service Fabric application

After this, we will select the **Stateless Service** ASP.NET Core, which will use the .NET Core runtime. As presented in the following figure, we will specify the service name and the location (optional) and then we click on **Create**.

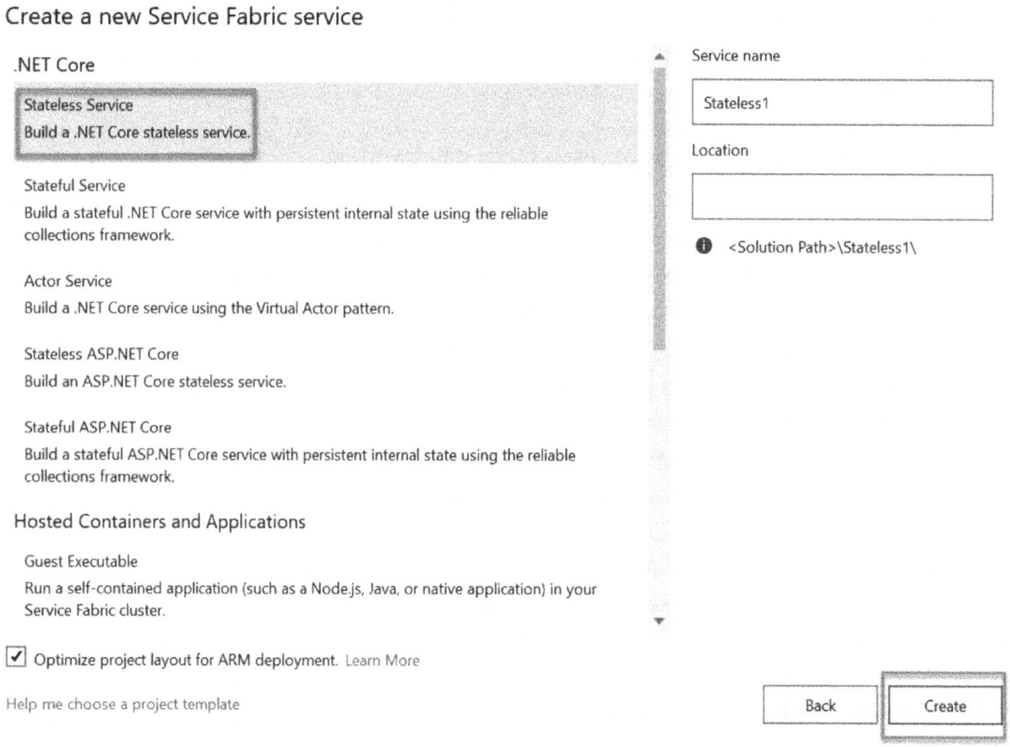

Figure 5.20 – Stateless Service on Azure Service Fabric

In the next window, we can select the application type. We will select **ASP.NET Core Web App (Model-View-Controller)**.

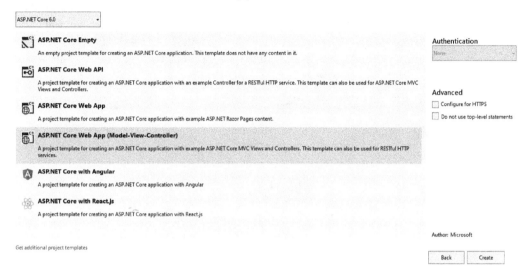

Figure 5.21 – Create a new ASP.NET Core web application with Service Fabric

We will explore the solution tree. We have two projects. The first project is the application itself and the second one is the service.

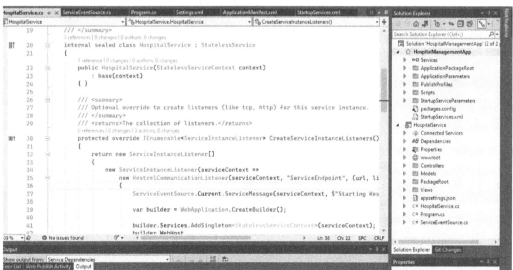

Figure 5.22 – The hospital management solution structure

In the application, you will find `.xml` files that represent metadata about the application. For example, inside `ApplicationPackageRoot`, we will find the `ApplicationManifest.xml` file, which represents the entire application because it includes the reference to all the services that constitute the application. In the `PackageRoot` folder inside the `HospitalService` project, you will find the `ServiceManifest.xml` file. Each service has its own manifest. The `ServiceManifest.xml` file contains a `ConfigPackage` element. Find the **Resources** element and the endpoints that this service will listen with. In this case, we only have a single endpoint.

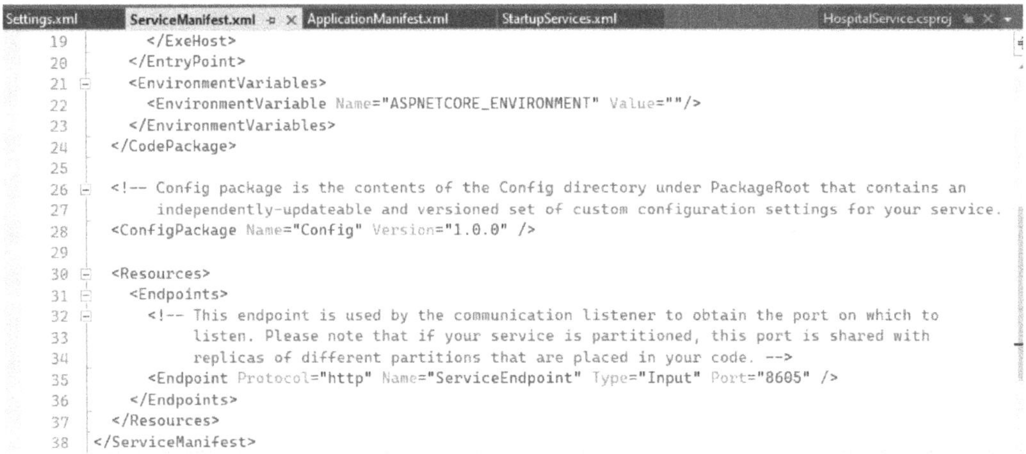

```
Settings.xml      ServiceManifest.xml  ⋆ ✕ ApplicationManifest.xml      StartupServices.xml                          HospitalService.csproj  ⋆ ✕  ▾
     19                </ExeHost>
     20              </EntryPoint>
     21              <EnvironmentVariables>
     22                <EnvironmentVariable Name="ASPNETCORE_ENVIRONMENT" Value=""/>
     23              </EnvironmentVariables>
     24            </CodePackage>
     25
     26            <!-- Config package is the contents of the Config directory under PackageRoot that contains an
     27                 independently-updateable and versioned set of custom configuration settings for your service.
     28            <ConfigPackage Name="Config" Version="1.0.0" />
     29
     30            <Resources>
     31              <Endpoints>
     32                <!-- This endpoint is used by the communication listener to obtain the port on which to
     33                     listen. Please note that if your service is partitioned, this port is shared with
     34                     replicas of different partitions that are placed in your code. -->
     35                <Endpoint Protocol="http" Name="ServiceEndpoint" Type="Input" Port="8605" />
     36              </Endpoints>
     37            </Resources>
     38          </ServiceManifest>
```

Figure 5.23 – ServiceManifest.xml

If we check our application in ASP.NET Core, in `Program.cs`, the service is registering itself to Service Fabric using this line of code:

```
ServiceRuntime.RegisterServiceAsync("HospitalServiceType",
                    context => new HospitalService(context)).
GetAwaiter().GetResult();
```

The `HospitalService` class is inheriting from `StatelessService`, which is one of the base classes in the Service Fabric API. We have two methods for listeners. The first one is `CreateServiceInstanceListeners`. It is returning a collection of listeners, `ServiceInstanceListener`, based on Kestrel. Kestrel is the web server included in .NET Core. In other words, this service is able to receive HTTP requests.

Deploying the application in a local cluster using Visual Studio

To deploy the application in a local cluster, we will use the **Publish** tool of Visual Studio. Right-click on the application with the **Service Fabric** icon and select **Publish**. In the target file, we will select the number of nodes configured in our cluster. In this case, we configured a single node in the connection endpoint. Ensure that your application will publish to a local cluster. In the end, the .xml files that include the application parameters and the startup service parameters are requested to finish the publishing process.

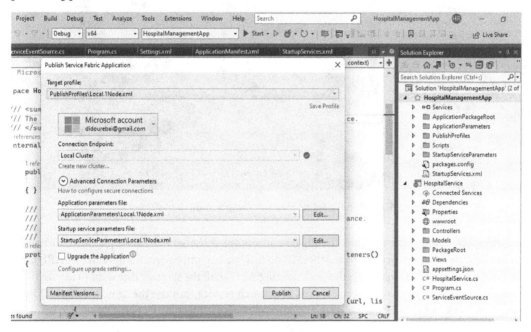

Figure 5.24 – Deploying a .NET application in a local Service Fabric cluster

Click on the **Publish** button, then we will check our application in Service Fabric Explorer.

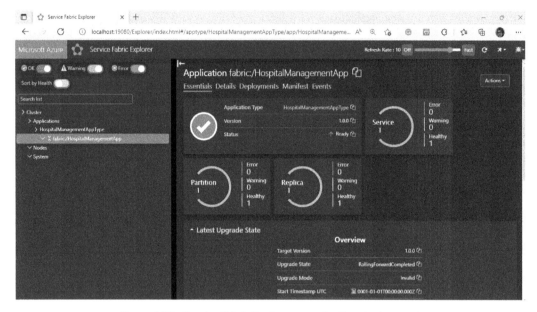

Figure 5.25 – Service Fabric Explorer – application deployment

A solution includes multiple services. To fully take advantage of the Service Fabric platform, we need to communicate from one service to another. To do this, we can implement an interface or API, for example. In this case, the interface is the public-facing API for all the services that want to invoke the hospital's functionality. Another service, such as a search service, can use the same interface. Both services are going to reference the interface. In a microservices architecture, it's perfectly valid to invoke one service directly from another. You can do this through the HTTP protocol and by exposing an API from the target microservice.

Exercise 5 – deploying an app to a Service Fabric managed cluster and containers

When the application is tested and deployed locally, we can publish it remotely to the Azure Service Fabric instance that we created previously. To do that, we will use the same publishing tool in Visual Studio.

Deploying an ASP.NET Core application to Azure Service Fabric

Right-click on the application, then select **Publish**. In the target profile, we will select `PublishProfiles\Cloud.xml`. We will select the account related to the Azure portal. In the connection endpoint, we can select an already created Azure Service Fabric cluster or you can create your cluster using Visual Studio. To do that, we will click on **Add Custom Endpoint**.

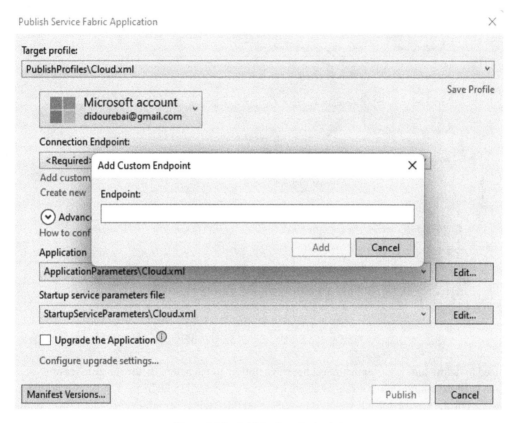

Figure 5.26 – Add Custom Endpoint

Next, we will add value to the endpoint. After following the previous steps to create the cluster, we will click on **Create a new Cluster**. The following figure shows the window for creating a Service Fabric cluster:

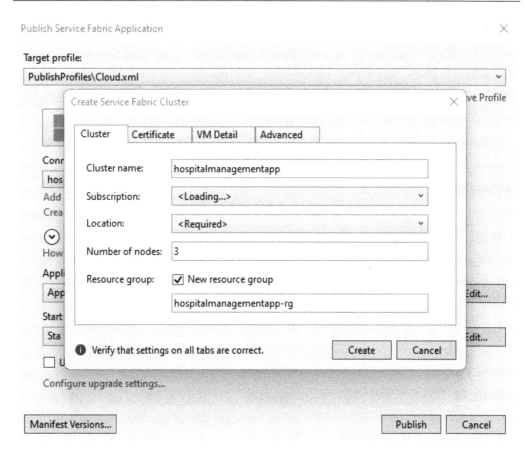

Figure 5.27 – Create Service Fabric Cluster

We need to define the cluster name, select the subscription and location, define the number of nodes, and select an existing resource group or create a new one. We need to fill in all of the information for every tab. For the **Certificate** tab, we need to add the certification information needed. In **VM Detail**, we will configure the VM that is used by the cluster. In **Advanced**, we need to add ports. After that, we will click on **Create**.

A new cluster will be created. At the bottom of the window, we need to select the application parameter files and the startup service parameter files that will be used for the cloud, so we will select Cloud. xml. This configuration ensures that each service is going to be deployed to all the nodes in the cluster. After this, we will click on **Finish**.

Building and executing a Docker container in Service Fabric

Another application model supported in Service Fabric is the use of Docker containers.

Containers are an ideal complement to the microservices architecture, thanks to the fact that they are isolated and lightweight processes.

Let's create a new Service Fabric application. First, we need to select a container template, set the service name, and specify the container image name. The image has to be reachable. The container image can be a public image in Docker Hub, for example, or a private image in Azure Container Registry or any private registry. In this case, we need to add a username and a password. You can also specify the host port and the container port that you want to use; for example, we can use port 80. Then, we will click on **Create**.

Figure 5.28 – Azure Service Fabric for containers

This type of application model doesn't add a service project since you won't write any code. This is because all the code is already inside the container. If we expand the service manifest for the package, the `ContainerHost` element contains the image name. Optionally, you can also set environment variables in the container using the `EnvironmentVariables` element. Locally, we can deploy this type of application in the local cluster. Remotely, you can deploy the application to Azure Service Fabric.

If you need to learn about creating a container, you can check out this book: *A Developer's Guide to Cloud Apps Using Microsoft Azure* by Packt Publishing.

Summary

In this chapter, we talked about the Azure Service Fabric platform. We explored the different elements of Service Fabric, such as clusters and nodes, the main difference between Service Fabric and Kubernetes, and we prepared a development environment to create and deploy an application.

We created a Service Fabric application using the Azure portal and the Azure CLI, we learned about scaling an Azure Service Fabric cluster, and we discovered more about the different models of Service Fabric template in Visual Studio 2022. We also created a .NET Service Fabric application, and we built and executed a Docker container in Service Fabric.

To publish our applications to Azure Service Fabric, we used the **Publish** tool in Visual Studio.

Applications use a database. In the next chapter, we will learn about the different classifications of data and how data is defined and stored. We will identify the characteristics of relational and non-relational data and the different Azure services for databases.

We will also explain the use of Azure Storage, Azure SQL Database, and Azure Cosmos DB, or a combination of them, to determine the operational needs, latency requirements, and transaction requirements of your data.

Questions

1. What is an Azure Service Fabric cluster?

2. What is the difference between Kubernetes and Service Fabric?

3. How do we deploy a containerized application on Azure Service Fabric?

6

Introduction to Application Data

Our ability to collect and process large amounts of data has grown over the years as technology has allowed us, as humans, to gather and collate significant amounts digitally. This has then enabled us to analyze large datasets from longer time periods and produce meaningful information that can improve or help to solve a problem and understand the cause.

In this chapter, we will cover the use of Azure Storage, Azure SQL Database, and Azure Cosmos DB, or a combination of them, to determine the operational needs, latency requirements, and transaction requirements for your data. We will present a microservices solution that includes some web APIs, and each API will use a different data service.

In this chapter, we're going to cover the following main topics:

- An overview of data classification and data concepts
- Exploring relational data concepts in Azure
- Exploring non-relational data concepts in Azure
- Exploring modern data warehouse analytics
- Getting started building with Power BI

An overview of data classification and data concepts

At some point in the design and implementation of your application, the developer or architect needs to determine the type, format, and location of the data to collect.

Data is a strategic, valuable asset. A holistic approach to data storage and data processing and an openness to new ideas can lead to incredible possibilities for taking applications to a new level and ensuring a stable and organized data stock.

Data is a set of collected information and facts, which can be numbers, character strings, and/or observations, to help us in decision-making. We can classify data into two classes, structured and semi-structured (unstructured):

- **Structured data**: This is tabular data that follows a schema, which implies that all data has the same properties, called fields. Structured data can be stored in a database table. To search for data inside a table that shares the same schema, we have query languages such as **Structured Query Language (SQL)**. A table is made up of rows and columns.

- **Unstructured data**: This is also called semi-structured data. It does not fit in tables, rows, and columns, and does not reside in a relational database, but still has some structure because it uses a data serialization language. Semi-structured data uses tags or keys that organize and provide a hierarchy for databases and graphs. We can also use the **JavaScript Object Notation (JSON)** format.

> **Important note**
> Semi-structured data is also known as non-relational or not only SQL (NoSQL) data.

Exploring relational data concepts in Azure

Relational database systems allow us to store and manage transactional and analytical data in organizations of all sizes around the world.

In a relational database, we model collections of entities from the real world as tables. Let's take the example of a solution for booking an appointment with a selected doctor. We might create a table for patients, doctors, and appointments. Every table will include rows, and each row will represent a single instance of an entity. Each row in a patient table contains the data for a single patient, each row in a doctor table defines a single doctor, and each row in an appointment table represents a patient that booked an appointment with a specific doctor. We will get more information related to the time and date of the appointment.

SQL is used to represent and search for data in a relational database. Common SQL statements used are SELECT, INSERT, UPDATE, DELETE, CREATE, and DROP to accomplish multiple tasks that need to be done in a database.

Microsoft Azure provides several services for relational databases, allowing you to run popular relational database management systems, such as SQL Server, PostgreSQL, and MySQL, in the cloud. You can select the relational database management system that best suits your needs and store your relational data in the cloud.

Most Azure database services are fully managed, which means that we save time by not having to manage databases. They have enterprise-class performance with built-in high availability that lets you scale quickly and achieve global distribution without caring about costly downtime. Developers can take advantage of industry-leading innovations such as built-in security with automatic threat detection and monitoring, which automatically adjusts to improve performance. In addition to all these features, you have guaranteed uptime.

When you decide to migrate your database to Azure or extend your on-premises SQL Server solution, you can move it to a virtual machine. We can use SQL Server for Azure virtual machines, for example. It allows you to run full instances of SQL Server in the cloud without having to manage on-premises hardware. However, this approach, known as lift and shift or infrastructure as a service, is not recommended because you have to manage your database. Azure offers a **Platform as a Service (PaaS)** that contains a series of **Database Management Systems (DBMSs)** managed by Microsoft in the cloud. Let's explore them!

We will start with **Azure SQL Managed Instance**. This is a **PaaS** option that is compatible with on-premises SQL Server instances while abstracting the underlying hardware and operating system. This service includes the automatic management of software updates, backups, and other maintenance tasks, reducing administrative costs when supporting database server instances.

We also have **Azure SQL Database**, which is a fully managed database designed for the cloud and is a highly scalable PaaS database service. You are able to create a managed database server in the cloud, and then deploy your databases on this server. This service contains the core database-level capabilities of SQL Server on-premises but not all its features. It is recommended to use it when building new cloud-oriented applications.

We have three options for Azure SQL Database:

- The first is a single database that allows you to set up and run a single SQL Server database.

- The second is an elastic pool. This is similar to a single instance – however, by default, several databases can share the same resources, such as memory, processing power, and data storage space.

- The last is a managed instance, which is a fully managed database instance designed for the easy migration of on-premises SQL databases.

These deployments have three different service tiers that you can choose from:

- **General Purpose**: Designed for standard workloads, this is the default tier.

- **Business Critical**: Designed for high-throughput applications and **online transaction processing (OLTP)**, this tier offers high resilience and the best performance with low latency.

- **Hyperscale**: This is an extension of the Business Critical tier. This tier is designed for large-scale OLTP deployments with autoscaling for compute resources such as compute and storage.

Now we have covered the different concepts of relational data in Azure, we'll move on to talking about **Azure SQL Edge**, which is an SQL engine that is optimized for **Internet of things (IoT)** scenarios that need to work with streaming time-series data.

Exploring non-relational data concepts in Azure

When we start building a new application, we need to think about how to store data. This usually takes the form of a relational database, where data is organized in linked tables and managed using SQL. However, many applications don't need the rigid structure of a relational database; we can use non-relational storage (commonly known as NoSQL).

Let's explore some characteristics of non-relational data. Non-relational data doesn't follow the rules of relational data. In its native form, data can be loaded quickly. If you have unknown data or queries, non-relational data will be more flexible and better than relational data, but it is less good for known data structures and known queries.

Entities have highly variable structures. For example, in a medical appointment database that stores information about patients, a patient can have more than a telephone number, landline, and mobile number. They can add multiple telephone numbers, such as a business number and another home phone number. Similarly, the addresses of patients might not always follow the same format. Addresses can be different according to states, regions, postal codes, and countries.

Let's discuss another scenario. If you ingest data, you want the process of capturing the data and saving it to be fast. When you want to process data and manipulate it in a set of rows in several tables of a relational database, at this stage, choosing a relational database is not appropriate because we have to transform the unstructured data into a compatible format before storing it in the database, which can cause latencies; even the search for information will be difficult to manage.

The entities will be stored in a set of collections or a container, not in a table like relational databases – but if we have two entities in the same collection, what happens? They will have a different set of fields, not like a set of columns in a relational table. The absence of a fixed schema leads to a description for each entity. For example, a non-relational collection of patient entities might look like this:

```
## Patient 1
ID: 1
Name: Hamida Rebai
Telephone: [ Home: 1-444-9999999, Business: 1-555-5555555,
Mobile: 1-444-4444444 ]
Address: [ Home: 2110 Rue De La Cantatrice Quebec, G2C0L6,
Canada,
Business: Marly Quebec G1X2Z2]
## Patient 2
ID: 2
Title: Mr
Name: Rayen Trabelsi
Telephone: [ Home: 0044-1999-333333, Mobile: 0044-17545-444444]
Address: [ FR: 86 High Street, Town X, County Y, 15000, FR,
US: 333 7th Street, Another Town, City A, CA, 11111]
```

In the previous example, we have fields that are preceded by a name. These fields can have one or more sub-fields, the same as for names. Brackets have been used to identify subfields of fields. If we want to add a new patient, we need to insert an entity with its fields labeled in a meaningful way. If an application queries this data, the application must have the ability to parse the information in the retrieved entity.

Entities inside a collection are usually stored in key-value format. The simplest type of non-relational database allows an application to specify a unique key or set of keys as a query condition. In the patient example, the database allows the application to query the patient by **identity** (**ID**) only.

To filter the data based on other fields, you need to scan the entire collection of entities and analyze each entity in turn, and then apply any query criteria to each entity to find a match. A query that retrieves patient details by identity can quickly identify the whole entity. A query that tries to find all customers with FR addresses must iterate over each entity and examine each field of each entity in turn. If the database includes 1 million entities, this query could take a long time to execute.

Now we have covered the different concepts of non-relational data in Azure, we'll move on to talking about modern data warehouse analytics.

Exploring modern data warehouse analytics

In the age of data mining, most organizations have multiple data stores, often with different structures and varying formats because we may need to collect data from multiple resources. They often have live incoming streams of data, such as sensor data in the case of IoT solutions and it can be expensive to analyze this data. There is often a wealth of useful information available outside the organization. This information could be combined with local data to add insights and enrich understanding. By combining local data with useful external information, it's often possible to gain insights into the data that weren't previously possible. The process of combining all of the local data sources is known as **data warehousing**. The process of analyzing streaming data and data from the internet is known as **big data analytics**. Azure Synapse Analytics combines data warehousing with big data analytics.

In this section, we will explore the different concepts of data warehousing. We will discuss the Azure data services to be used for modern data warehouses and the data warehouse architecture and workload.

Exploring data warehousing concepts

A modern data warehouse might contain a mixture of relational and non-relational data, including files, social media streams, and IoT sensor data. Azure provides a collection of services you can use to build a data warehouse solution, such as **Azure Data Factory** (**ADF**), Azure Databricks, Azure Synapse Analytics, Azure Data Lake Storage, and Azure Analysis Services. You can use tools such as Power BI to analyze and visualize the data, generating reports, charts, and dashboards.

Any modern data warehouse solution should be able to provide access to raw data streams and business insights from that data.

Azure data services for modern data warehouses

In this section, we will further discuss the data services provided by Azure. These services allow you to combine data from multiple sources, reformat the data into analytic models, and save those models for later queries, reports, and visualizations.

To implement a modern data warehouse using Azure database services, we will use these services:

- **ADF** is an Azure cloud-based data integration PaaS solution that allows you to orchestrate and automate data movement and data transformation. Data integration is the process of combining data from multiple sources and providing an integrated view of it.

- **Azure Data Lake** is built on Azure Blob Storage, Microsoft's object storage solution for the cloud. This solution features cost-effective tiered storage and high availability/disaster recovery capabilities. It integrates with other Azure services such as ADF, a tool for creating and running **extract, transform, load** (ETL), and **extract, load, transform** (ELT) processes.

- **Azure Databricks** is a cloud-based platform that leverages Spark for big data processing with built-in SQL database semantics and management interfaces for rich data analytics and machine learning. It is a cloud-integrated version of Azure for the Databricks platform. With Databricks, we are able to use existing Databricks and Spark skills to create an analytical data store. We can also use Azure Databricks' native notebook support to query and visualize data in an easy-to-use web-based interface.

- **Azure Synapse Analytics** is an integrated end-to-end solution for data analytics at scale. It includes multiple technologies and capabilities to combine the data integrity and reliability of a scalable, high-performance SQL Server-based relational data warehouse with the flexibility of a data lake and open source Apache Spark. It also contains native support for log and telemetry analytics using Azure Synapse Data Explorer pools and built-in data pipelines for data ingestion and transformation.

- **Azure Analysis Services** is a fully managed PaaS. It delivers cloud-oriented enterprise-grade data models. In Azure Analysis Services, we can use advanced mashup and modeling capabilities in order to combine data from multiple data sources into a single trusted semantic tabular data model, define metrics, and secure our data.

- **Azure HDInsight** is an enterprise-class, open source, managed full-spectrum analytics service in the cloud. It is a cloud distribution of Hadoop components. With HDInsight, you can use open source frameworks such as Apache Spark, Apache Hive, LLAP, Apache Kafka, and Hadoop in your Azure environment.

- **Microsoft Purview** is an enterprise-wide data governance and discoverability solution, with features such as the ability to create a map of data and track the data lineage across different data sources and systems so you can find reliable data for reporting and analytics. In order to enforce data governance across an organization, we can use Microsoft Purview. Microsoft Purview ensures the integrity of the data used to support analytical workloads.

Getting started building with Power BI

Microsoft Power BI is a collection of software services, applications, and connectors. They work together to transform unrelated data sources into coherent information, in a visually immersive and, above all, interactive way, facilitating the exploitation of the data afterward.

Data can be diverse – for example, a Microsoft Excel data workbook or a collection of cloud-based and on-premises hybrid data warehouses. Power BI makes it easy to connect to these data sources, view (or exploit) what is important, and share it according to permissions or for everyone by being public.

Power BI has a Microsoft Windows desktop application called Power BI Desktop, an online **Software-as-a-Service** (**SaaS**) service called the Power BI service, but another mobile version is available on any device using the native BI mobile application for Windows, iOS, and Android apps.

In the next section, we will see how to install and configure Power BI Desktop.

Power BI Desktop

To install Power BI Desktop, you can follow this link to download it from Windows Store: `https://aka.ms/pbidesktopstore`. Or, you can download it using this link: `https://go.microsoft.com/fwlink/?LinkID=521662`. You must manually update it periodically:

Figure 6.1 – Power BI Desktop interface

Before signing in to Power BI, you need to create an account. Go to this link and use your email address: `https://app.powerbi.com/`. If you need more details related to the Power BI service, you can follow this documentation link: `https://learn.microsoft.com/en-us/power-bi/consumer/end-user-sign-in`.

Summary

In this chapter, we discussed data classification and different data concepts. We explored relational data and non-relational data concepts in Azure. We explored modern data warehouse analytics and, at the end of the chapter, we learned about Power BI.

In the next chapter, we will learn how to provision and deploy Azure SQL Database and Azure SQL Managed Instance. We are able to select from multiple options when performing a migration to the SQL PaaS platform.

7

Working with Azure SQL Database

In this chapter, we will learn how to provision and deploy Azure SQL Database and Azure SQL Managed Instance. We will also learn how to choose an option from multiple options when performing a migration to the SQL **platform as a service (PaaS)** platform. A web API will store data on Azure SQL Database.

In this chapter, we're going to cover the following main topics:

- Exploring PaaS options for deploying SQL Server in Azure
- Exercise 1 – deploying a single SQL database
- Exercise 2 – deploying Azure SQL Database elastic pools
- Exercise 3 – deploying SQL Managed Instance
- Exercise 4 – connecting Azure SQL Database to an ASP.NET app

Exploring PaaS options for deploying SQL Server in Azure

In this section, we will learn about the different PaaS options to deploy SQL Server in Azure. If we use SQL Server as a **relational database management system (RDBMS)**, we want to migrate our database without having to make any conversions or thinking about reinstalling SQL Server and configuring it again. Azure offers a PaaS that provides a complete development and also deployment environment that is easy to use for cloud-oriented applications.

There are a couple of SQL Server options within Azure, such as SQL Server on Azure Virtual Machines or SQL Managed Instances.

Azure SQL Database

Azure SQL Database is a hosted SQL database service in Azure. It runs on the SQL Server database engine. There are some important differences between Azure SQL Database and the traditional SQL Service. But most database administrators using SQL Server are able to migrate to Azure SQL Database. Azure SQL Database makes it extremely easy to scale a database. We are able to replicate a database in one or more other locations around the world, which can improve performance if your application is used worldwide.

Azure SQL Database is available not only as a PaaS but also as serverless. The difference between them is that the PaaS option requires you to select a level of performance, and you will pay that fixed rate per hour for access to the database. You can easily scale to a bigger or smaller performance plan without causing disruption to your applications or your users by selecting the option within the Azure portal. The serverless option allows you to pay only for usage. In fact, the database shuts down when you're not using it. This can be ideal for underused databases, such as if you have a developer-only tool that needs a database to run but doesn't need to run 24/7 and can be available immediately. Azure offers 99.99% reliability for Azure SQL Database and its default configuration.

Azure offers a high-level tier if you have a database that is an important resource for your business. This is sometimes referred to as the premium tier in the **database transaction unit** (**DTU**) pricing model. Business-critical SQL databases use a cluster of SQL database processes to respond to database requests. Having multiple SQL engine processes makes your database more reliable because the failure of one process doesn't affect you directly.

Note that the Azure SLA doesn't grow between **Standard** and **Business Critical**, although you can expect higher availability in practice. But you can further increase your guaranteed uptime SLA by adding zone-redundancy deployments. You can only do enterprise-critical zone redundancy deployments, which brings you to 99.995% availability. Zone-redundant deployments involve deploying multiple replicas of a database that are physically separated from each other in availability zones.

Let's discover Azure SQL Managed Instance in the next section.

Azure SQL Managed Instance

We are able to move SQL Server to Azure using two different options. **Infrastructure as a service** (**IaaS**) allows you to create an Azure SQL Server virtual machine and PaaS allows you to create a fully managed Azure SQL database. There is another option that sits right in between these two and it's called **SQL Managed Instance**. SQL Managed Instance combines the benefits of both of the two alternatives (PaaS and IaaS). Like a virtual machine, managed instances have almost 100% compatibility with on-premises SQL Server. This includes **SQL Server Agent** support, access to a temporary database, cross-database queries, distributed transactions, and common language runtimes. They are not available in a fully managed Azure SQL database. A managed instance also supports up to 100 individual databases in the same instance. But unlike virtual machines, managed instances also support automatic backup scheduling, automatic remediation, and built-in high availability just like Azure SQL Database.

Creating an Azure SQL Database instance

To create a new SQL database or SQL Managed Instance or only migrate to a SQL virtual machine, we will create an Azure SQL instance that includes all these options. To do that, follow these steps:

1. Open the Azure portal and select **Create a new resource**.

2. Search for Azure SQL, then click on **Create**, as presented in the following figure:

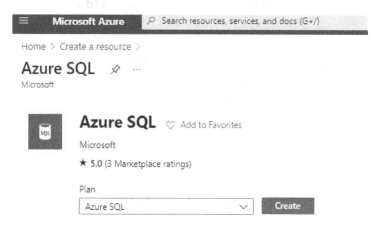

Figure 7.1 – Azure SQL service

Or, use this link, https://portal.azure.com/#create/Microsoft.AzureSQL, to display the **Select SQL deployment option** page, as presented in the following figure:

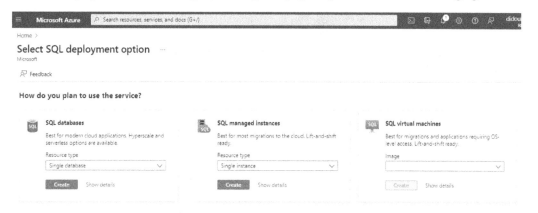

Figure 7.2 – Select SQL deployment option page

Every service is available in two different deployment models. We can create a single database for all services or we can select a different option. For example, for SQL database services, we can also select **Elastic pool** or **Database Server**. For SQL Managed Instances, we have a single database in Azure Arc, and in SQL virtual machines, we can select the SQL Server image to install it.

Let's now discover how to create and deploy a single SQL database in the next section.

Exercise 1 – deploying a single SQL database

In a single database, each database is isolated; so, every database is billed and managed independently in terms of scope, scale, and data size. We have two models; the first one is a DTU-based purchasing model that provides a set of prices versus the performance packages to select for easy configuration, and the second one is a vCore-based purchasing model that offers a wide range of configuration controls and also provides **Hyperscale** and **Serverless** options to automatically scale your database according to your needs and the required resources. In both models, each database has its own dedicated resources with its own service tier, even if they are deployed on the same logical server. This is why it is considered a deployment model for Azure SQL Database.

We will start by creating a SQL database. We'll use the same page as shown in *Figure 7.2* and leave **Resource type** set to **Single database**. Then, click on the **Create** button. Let's start with the **Basics** tab:

1. In **Project details**, select the subscription and the resource group (we can create a new one). In the following figure, we can see the **Project details** part:

Figure 7.3 – Create SQL Database – Project details part

2. In **Database details**, we fill in the different required settings for a database. We need to select an existing SQL database server but we can create it before completing the different settings related to the database. In the following figure, we can see the **Database details** part:

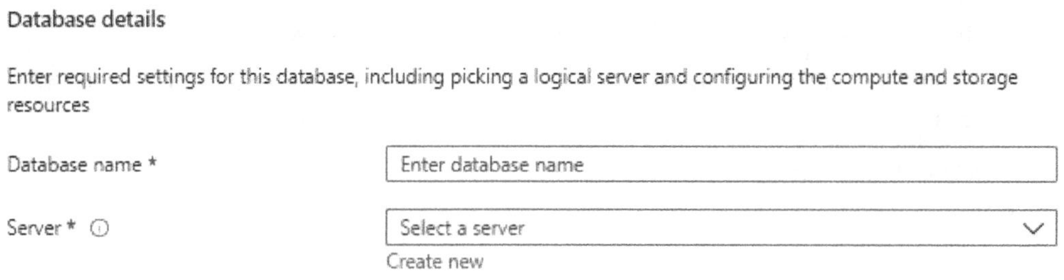

Database details

Enter required settings for this database, including picking a logical server and configuring the compute and storage resources

Database name *   Enter database name

Server * ⓘ   Select a server   ⌄
                                   Create new

Figure 7.4 – Create SQL Database – Database details part

3. If you don't have a SQL database server, you will have to create a new server. Fill out the different settings as follows:

    A. **Server name**: This should be unique, not just within a subscription but across all Azure servers.

    B. **Location**: Select a location from the drop-down list. In the following figure, we can see the first part, which is related to the details of a new SQL database server:

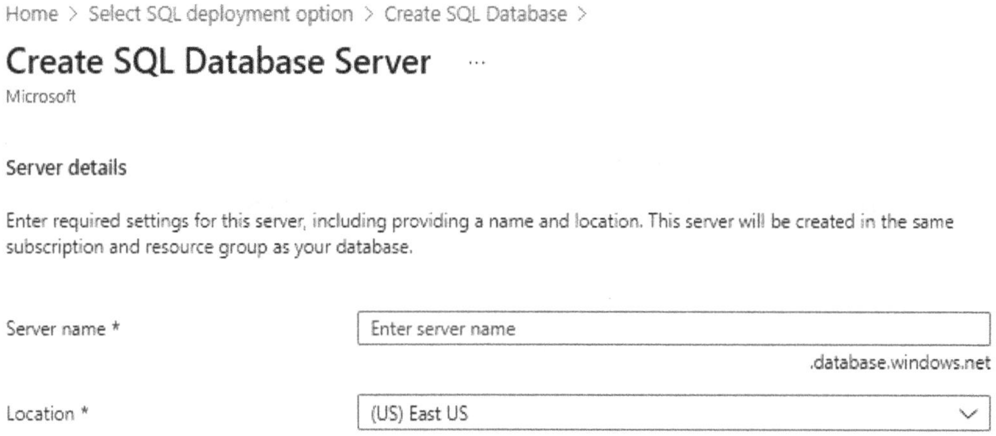

Home > Select SQL deployment option > Create SQL Database >

Create SQL Database Server   ⋯
Microsoft

Server details

Enter required settings for this server, including providing a name and location. This server will be created in the same subscription and resource group as your database.

Server name *   Enter server name
                                                                      .database.windows.net

Location *   (US) East US   ⌄

Figure 7.5 – Create SQL Database Server – Server details

    C. **Authentication method**: We have three options to select:

    • **Use only Azure Active Directory (Azure AD) authentication**

    • **Use both SQL and Azure AD authentication**

    • **Use SQL authentication**

        Select **Use SQL authentication**.

D. **Server admin login**: This is the administrator login of the server.

E. **Password**: The password should meet the requirements, and we need to confirm the password in another field. Click on **Ok** to complete the creation of the server.

Select **OK** to go back to the previous page.

In the following figure, we can see the **Authentication** part when we create a SQL database server:

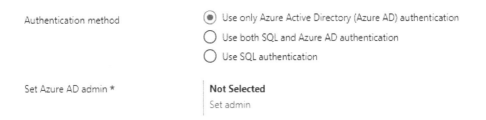

Figure 7.6 – Create SQL Database Server – Authentication part

4. Select **No** for the question **Want to use SQL elastic pool?**.

5. For **Workload environment**, choose between **Development** and **Production**.

    These options are shown in the following figure:

Figure 7.7 – Create SQL Database – disable SQL elastic pool and workload environment

6. Under **Compute + storage**, we need to select **Configure database**:

Figure 7.8 – Configure database

We will see the options shown in the following figure:

Service and compute tier

Select from the available tiers based on the needs of your workload. The vCore model provides a wide range of configuration controls and offers Hyperscale and Serverless to automatically scale your database based on your workload needs. Alternately, the DTU model provides set price/performance packages to choose from for easy configuration. Learn more ◻

| Service tier | General Purpose (Scalable compute and storage options) ⌄ |
| | Compare service tiers ◻ |

Compute tier
- ⦿ **Provisioned** - Compute resources are pre-allocated. Billed per hour based on vCores configured.
- ◯ **Serverless** - Compute resources are auto-scaled. Billed per second based on vCores used.

Compute Hardware

Select the hardware configuration based on your workload requirements. Availability of compute optimized, memory optimized, and confidential computing hardware depends on the region, service tier, and compute tier.

Hardware Configuration
Standard-series (Gen5)
up to 80 vCores, up to 415.23 GB memory
Change configuration

Cost summary

General Purpose (GP_Gen5_2)
Cost per vCore (in EUR) ¹   93.71
vCores selected   x 2
Cost per GB (in EUR)   0.10
Max storage selected (in GB)   x 41.6
ESTIMATED COST / MONTH   191.45 EUR
NOTES

Figure 7.9 – Select service and compute tier

In this window, select the available tiers according to the needs of your workload. We can leave **Service tier** as **General Purpose** or we can select another option, such as **Hyperscale** or **Business Critical**, which are related to the vCore-based purchasing model, or **Basic**, **Standard**, or **Premium** for a DTU-based purchasing model.

In **Compute tier**, select **Serverless** to ensure the auto-scaling of the compute resources and that they are billed per second.

If we move to **Compute Hardware**, we can change the configuration of the hardware configuration. By default, it is **Gen5**, but we have more options. We will leave the default configuration.

Next, update the maximum and minimum number of vCores. By default, we have **2** for the maximum and **0.5** for the minimum, as presented in the following figure. We have more options to update, such as **Auto-pause delay**, which can be used to pause a database automatically if it is not active in a specific period.

Max vCores

◯   2

Min vCores

◯   0.5 vCores

2.02 GB MIN MEMORY   3 GB MAX MEMORY

Figure 7.10 – Maximum and minimum vCores configuration for compute hardware

At the end, we have to choose **Yes** or **No** for database zone redundancy. After that, click on **Apply**.

7. Select the backup storage redundancy. We have three options. Select **Locally-redundant backup storage**.

8. Move to the **Networking** tab. In this tab, configure the network access and connectivity for the server. For **Connectivity method**, select **Public endpoint**. For **Firewall rules**, set **Add current client IP address** to **Yes**. Leave **Allow Azure services and resources to access this server** set to **Yes** to be able to connect to the database from our local machine, as presented in the following figure:

Network connectivity

Choose an option for configuring connectivity to your server via public endpoint or private endpoint. Choosing no access creates with defaults and you can configure connection method after server creation. Learn more ☐

Firewall rules

Setting 'Allow Azure services and resources to access this server' to Yes allows communications from all resources inside the Azure boundary, that may or may not be part of your subscription. Learn more ☐
Setting 'Add current client IP address' to Yes will add an entry for your client IP address to the server firewall.

Figure 7.11 – Configure network access in the Networking tab

9. Under **Connection policy**, we will leave the default connection policy, and we will leave the **Minimum TLS** version at the default of TLS 1.2.

10. On the **Security** tab, you can choose to start a free trial of **Microsoft Defender for SQL**, as well as configure **Ledger**, **Managed identities**, and **transparent data encryption** (TDE) if you want.

11. In **Additional settings | Data source | Use existing data**, select **Sample** to start testing the service, **Backup** to select an existing backup, or **None** if you want an empty database to load after configuration:

Figure 7.12 – Configure the Additional settings tab in Azure SQL Database

If we select **Sample**, this will create an **AdventureWorksLT** sample database so there are some tables and data to query and experiment with. We can configure a database collection that defines the different rules that will sort and compare the data.

12. Click, in the end, on **Review + Create**, and after reviewing, select **Create**.

Once the database is created, we will click on **Go to resource**.

In **Overview**, we have the different configurations set during creation.

Select **Getting started** to explore the different options to start working with the database. We can configure the access to SQL Server, connect to the application using the connection strings, or start developing using Azure Data Studio, where we can manage our data in the database:

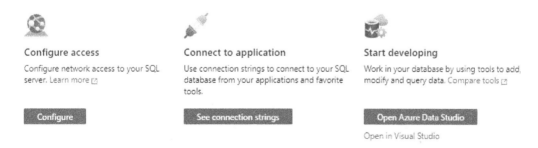

Figure 7.13 – Getting started with Azure SQL Database

We can test the database by using **Query Editor**. We already have a sample database. You can enter any query in the **Query Editor** pane.

We can create a single database using the Azure CLI or PowerShell.

Sometimes, we need to manage multiple databases. In this case, we will use Azure SQL Database elastic pools. We will see in the next section how we can deploy them in Azure.

Exercise 2 – deploying Azure SQL Database elastic pools

If we want to manage and scale multiple databases that have various and unpredictable usage requirements, we will use Azure SQL Database elastic pools.

Pools are suitable for large numbers of databases with specific usage patterns. For a given database, this pattern is characterized by infrequent usage peaks and low average usage. Otherwise, do not place multiple databases under moderate sustained load in the same elastic pool.

Pools simplify management tasks by running scripts in elastic jobs. Elastic jobs eliminate most of the tedious work associated with numerous databases.

Pooled databases generally support the same business continuity features available with single databases.

Azure SQL Database elastic pools are a simple and cost-effective solution. They are a deployment option, which means we purchase Azure compute resources (eDTUs) and storage resources to be shared between all the included databases. Why? Because the databases in an elastic pool reside on a single server and share a fixed number of resources for a fixed price. **Software as a service (SaaS)** developers can optimize the pricing performance of a group of databases within a well-defined budget by benefiting from performance elasticity for each database, by using the elastic pools in Azure SQL Database. In an elastic pool, we are able to add databases to the pool and set the maximum and minimum resources (DTUs or vCore) for the databases according to our budget.

We will follow these steps to create a SQL database elastic pool:

1. Open the Azure portal, select **Create a resource**, and search for SQL Elastic database pool. You will see the following screen:

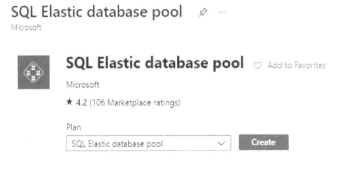

Figure 7.14 – Create SQL elastic database pool

2. In the **Basics** tab, fill in all the information needed to create a SQL elastic pool. Under **Project details**, select the subscription and the resource group. Under **Elastic pool details**, we will introduce the required settings for this pool, select a logical server, and then configure compute and storage resources. We have the following screen to add the elastic pool name and server:

Elastic pool details

Enter required settings for this pool, including picking a logical server and configuring the compute and storage resources.

Elastic Pool Name *	rebaipool
Server * ⓘ	rebaiserver (East US)
	Create new
Compute + storage * ⓘ	**GeneralPurpose** Gen5, 2 vCores, 32 GB, 0 databases Configure elastic pool

Figure 7.15 – Elastic pool details in a SQL elastic pool

3. In **Compute + storage**, we need to configure the elastic pool to select multiple databases. Click on **Configure elastic pool**. We have the **Pool settings** tab, which includes different settings such as the service tier, the compute hardware, and the number of vCores. We can enable elastic pool zone redundancy. In the **Databases** tab, we will add the database. Click on **Add databases**. Another window will be displayed, which will include all single databases. You can select one or more databases to add to the elastic pool. They will share the same compute resources:

Figure 7.16 – Add databases to the elastic pool

4. Click on **Apply**, then select **Review + create**, then **Create**, and the elastic pool will be created.

5. Once created, select **Go to resource**, and in **Overview**, we can check the resource consumption and identify which database is consuming the most resources. It is a monitoring tool that helps us to run diagnostics on the performance issues of each database. We can also use the **Monitoring** feature, under **Database Resource Utilization**, to display the performance of every database within the pool.

6. We can also configure our elastic pool by selecting **Configure**:

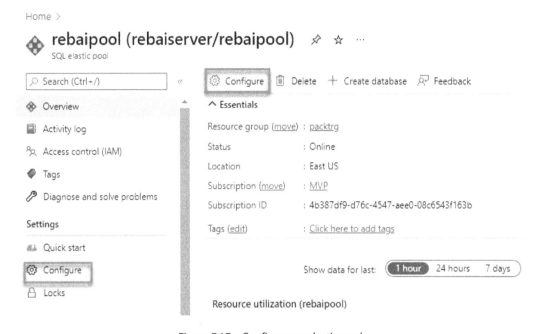

Figure 7.17 – Configure an elastic pool

7. If we need to update the elastic pool configuration by increasing or decreasing the allocated resources, we can adjust the pool size, the service tier, or the resources of every database. We can add or remove the databases inside the pool:

Figure 7.18 – Pool settings in an elastic pool

If you have a large number of databases to migrate using a specific pattern, Azure SQL Database elastic pools are the best solution. In the next section, we will deploy a SQL Managed Instance.

Exercise 3 – deploying SQL Managed Instance

If we go back to the SQL deployment page shown in *Figure 7.2*, we can now select **SQL managed instances**. Leave **Resource type** as **Single instance** and click on **Create**:

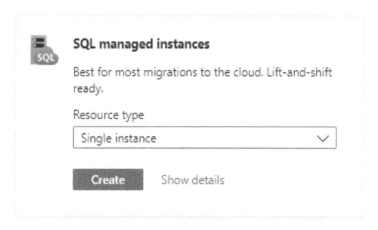

Figure 7.19 – Create a SQL managed instance

In the **Basics** tab, we will introduce the information related to the project details, the managed instance details, and authentication; it is similar to a SQL database.

In the **Networking** tab, we will configure the virtual network and public endpoint. You can leave the default configuration; it depends on your internal configuration.

In the **Security** tab, we need to enable Microsoft Defender for SQL. We can configure system-assigned and user-assigned managed identities to enable central access management between this database and other Azure resources.

We have more tabs, such as **Additional settings**, which are used to customize additional configuration parameters, including geo-replication, time zone, and collation.

Then, we click on the **Review + create** button.

Azure SQL Managed Instance is used for an easy lift and shift because it is compatible with on-premises servers. It is a fully managed PaaS and easy to use and configure for modern databases.

Let's explore how we can connect Azure SQL Database to an ASP.NET application.

Exercise 4 – connecting Azure SQL Database to an ASP.NET app

We will learn in this section how to connect an ASP.NET application to SQL Database.

The application is already using SQL Server 2019. We start by migrating the database to SQL Database using the Data Migration Assistant migration tool (to download it, use this link: `https://www.microsoft.com/en-us/download/details.aspx?id=53595`).

We will start by configuring the database connection, then we will update the connection string in ASP. NET Core. The application is using Entity Framework to connect and manage Azure SQL Database.

Creating and configuring the database connection

We will follow these steps to create a database connection:

1. Open Visual Studio 2022 and use **SQL Server Object Explorer**, as presented in the following figure:

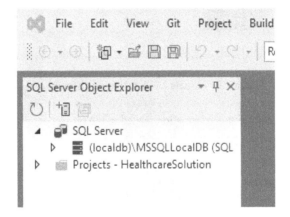

Figure 7.20 – SQL Server Object Explorer

2. At the top of SQL Server Object Explorer, right-click on **SQL Server**, then select **Add SQL Server**, and a new dialog will be displayed, as presented in the following figure:

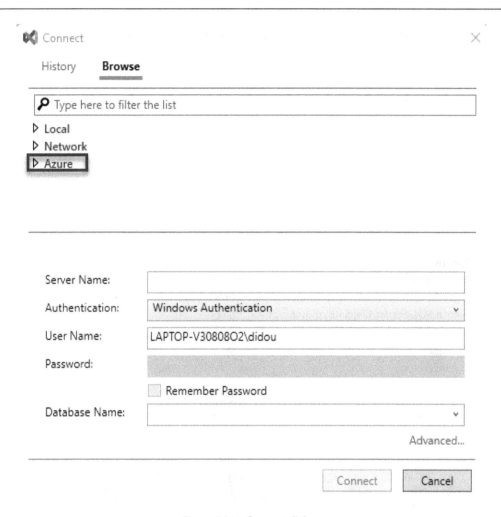

Figure 7.21 – Connect dialog

We will follow these steps to configure the database connection:

1. In the **Connect** dialog, expand the **Azure** node. All Azure SQL Database instances will be listed.

2. We will select the database that we will use with our application.

3. The information related to the database will be filled in at the bottom, but we have to enter the administrator password after we click on the **Connect** button.

4. The **Create a new firewall rule** dialog is displayed. To connect to your database from outside Azure, you have to create a server-level firewall rule. A firewall rule allows the public IP address of the local computer.

In the following figure, we configure a firewall rule by allowing the internet IP to be able to access the resource from Visual Studio:

Figure 7.22 – Allow public internet IP addresses in the Azure portal

5. When the firewall settings for Azure SQL Database have been configured, we will select the database in SQL Server Object Explorer. Right-click on the server's name and select **Properties** and get the connection string from there. In the following figure, we have displayed the database server in SQL Server Object Explorer:

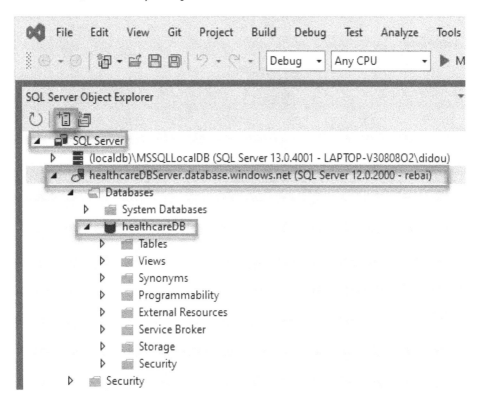

Figure 7.23 – Display the database server in SQL Server Object Explorer

6. Go back to our application. Open the `appsettings.json` file and add the connection string:

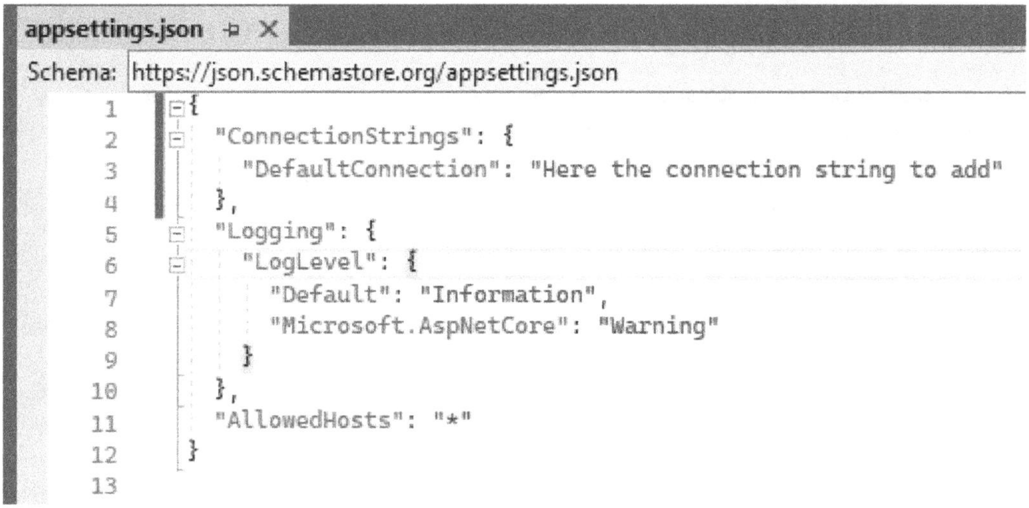

```
appsettings.json ⊕ ✕
Schema: https://json.schemastore.org/appsettings.json
  1  {
  2      "ConnectionStrings": {
  3          "DefaultConnection": "Here the connection string to add"
  4      },
  5      "Logging": {
  6          "LogLevel": {
  7              "Default": "Information",
  8              "Microsoft.AspNetCore": "Warning"
  9          }
 10      },
 11      "AllowedHosts": "*"
 12  }
 13
```

Figure 7.24 – Connection string of the database in Visual Studio 2022

7. We can use the `Database.Migrate()` call to help us run the database in Azure SQL Database.

If you use a local database for development and Azure SQL Database for production, you need to configure each environment in `Program.cs`.

Summary

In this chapter, we explored the different services to deploy SQL Server in Azure. We started by deploying a single SQL database. We have multiple options, such as the serverless database option, that permit us to create a low-cost database. Serverless is a compute tier for single databases in Azure SQL Database that automatically scales compute based on workload demand and bills for the amount of compute used per second. After we deployed the **Azure SQL Database** elastic pool, which includes the serverless option, we deployed a SQL Managed Instance and, in the end, we connected an Azure SQL database to an existing ASP.NET application.

In the next chapter, we will talk about the different storage options that are available in the Azure Storage services, and the scenarios in which each storage option is appropriate.

Further reading

If you want more information related to the Azure database service, you can check out these links:

- `https://learn.microsoft.com/en-us/azure/azure-sql/database/sql-database-paas-overview?view=azuresql`

- `https://learn.microsoft.com/en-us/azure/azure-sql/database/?view=azuresql`

Questions

1. What is Azure SQL Database?
2. How do we connect Azure SQL Database to an ASP.NET app?

8

Working with Azure Storage

Azure Storage is a Microsoft cloud storage solution for more than just data storage. Azure Storage offers a scalable and secure object store for a variety of data objects, a filesystem service in the cloud, a messaging store for relevant messaging, and an SQL store.

This chapter will cover the different storage options that are available in Azure Storage and the scenarios in which each storage option is appropriate.

In this chapter, we're going to cover the following main topics:

- Azure Storage account
- Exploring Azure Table storage
- Exploring Azure Blob Storage
- Exploring Azure Disk Storage
- Exploring Azure Files

Azure Storage account

In the healthcare solution, we have a platform for patient medical records. It includes unstructured data such as documents, messages, notes, and images related to a patient, for example, diagnoses documents, annual exams, and disease surveillance. All these documents are stored on a virtual machine using a custom solution to extract or load the unstructured data.

Azure Storage is a service that can be used to store files, messages, tables, and other types of data. It's scalable, secure, and easy to manage and can be accessed from anywhere. Azure Storage is following two cloud models: infrastructure as a service by using virtual machines and platform as a service by using native cloud services. Azure Storage includes four data services, which can be accessed through a storage account.

Each type supports different features and has a unique URL address to access. The data and your Azure Storage account are always replicated to ensure durability and high availability.

To create an Azure Storage account in Azure, the first step consists of using the Azure portal, PowerShell, or the Azure **Command-Line Interface (CLI)**.

If you're using the Azure portal, select **Create a resource**, then **Storage** as the category, and then **Create** under **Storage account**:

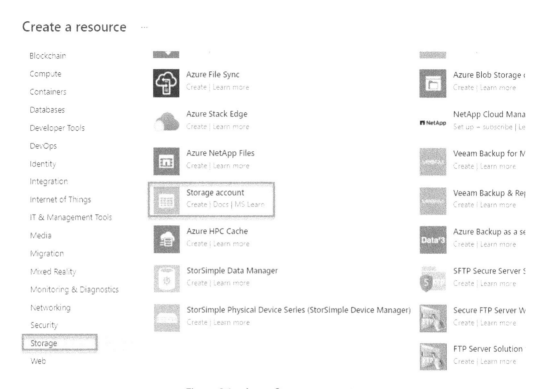

Figure 8.1 – Azure Storage account

In the **Basics** tab, we will introduce the project details. This part is the same for all Azure services. We have the instance details, which include the storage account name and the location. For **Performance**, we need to select **Standard** or **Premium**. The following screenshot shows the project details:

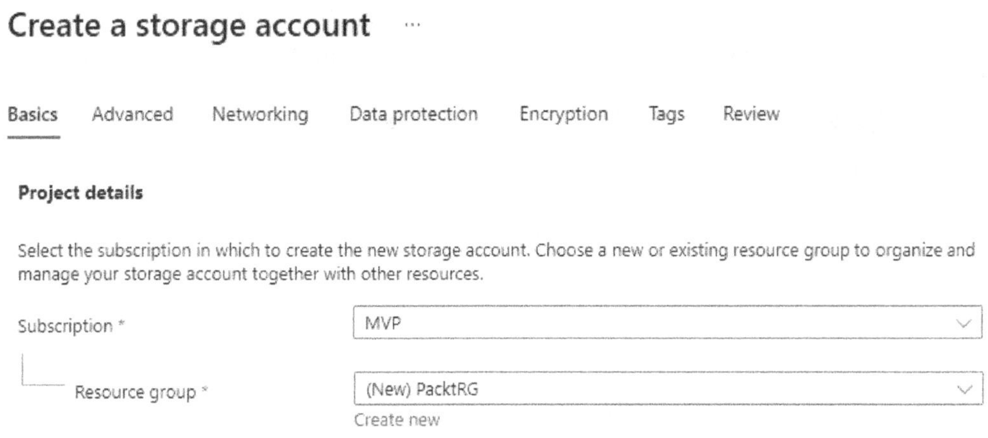

Create a storage account   ...

Basics   Advanced   Networking   Data protection   Encryption   Tags   Review

Project details

Select the subscription in which to create the new storage account. Choose a new or existing resource group to organize and manage your storage account together with other resources.

Subscription * | MVP |

Resource group * | (New) PacktRG |
Create new

Figure 8.2 – Create a storage account – Project details in the Basics tab

In the following screenshot, you can see how we configure the instance details:

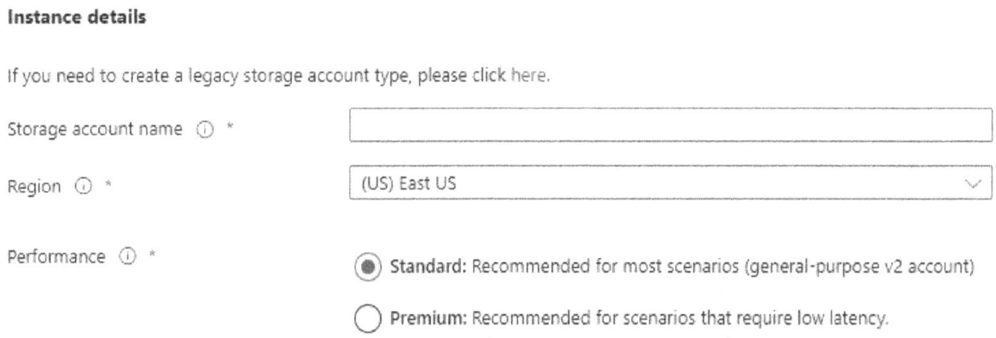

Instance details

If you need to create a legacy storage account type, please click here.

Storage account name ⓘ * | |

Region ⓘ * | (US) East US |

Performance ⓘ *   ⦿ Standard: Recommended for most scenarios (general-purpose v2 account)

◯ Premium: Recommended for scenarios that require low latency.

Figure 8.3 – Create a storage account – Instance details in the Basics tab

The data in the Azure storage account is always replicated to ensure durability and high availability. Select a replication strategy that matches your durability requirements. Select from the list of four elements:

- **Locally-redundant storage (LRS)**: This is the lowest-cost option. It includes basic features and protection against server rack and drive failures. You can select this option if you have a simple scenario, which is not critical or for testing.

- **Geo-redundant storage (GRS)**: This is used in backup scenarios because it is an intermediate option with a failover capability in another secondary region.

- **Zone-redundant storage (ZRS)**: This is also an intermediate option, but it provides protection against data center-level failures. It can be used for high-availability scenarios.

- **Geo-zone-redundant storage (GZRS)**: This is the best option for critical data scenarios because it offers optimal data protection, including a combination of the features of GRS and ZRS.

In the following figure, you can see the options for configuring redundancy:

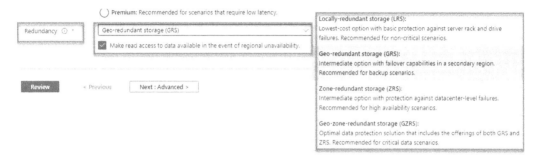

Figure 8.4 – Create a storage account – Redundancy configuration

Here, we have the classic deployment model, but if we need to create a legacy storage account type in **Instance details**, we will click on **here**, as presented in the following figure:

Instance details

If you need to create a legacy storage account type, please click here.

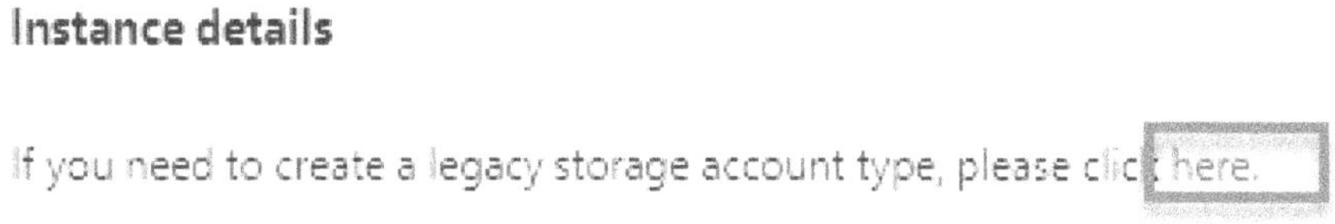

Figure 8.5 – Select a legacy storage type

A new line will be added that is related to the storage account. If you need more details, you can follow this link: `https://docs.microsoft.com/en-ca/azure/storage/common/storage-account-create?tabs=azure-portal`. This is used to choose an account type that matches your storage needs and optimizes your costs.

A storage account contains all Azure Storage data objects, such as blobs, files, and disks.

In the next section, we will start exploring Azure Table Storage.

Exploring Azure Table Storage

Azure Table Storage provides support for storing structured data. It implements a NoSQL key-value model, which means that it doesn't have any concept of relationships, stored procedures, secondary indexes, or foreign keys.

Azure Table Storage is not the only service you can use to work with semi-structured data. So, why use this service instead of other options? Let's start by listing the benefits:

- Azure Table Storage can store petabytes of data at a low cost, which is really important when it comes to building highly scalable applications where thousands or even millions of users interact with them.

- Azure Table Storage also supports a flexible data schema, which is excellent for flexible datasets, such as web applications, user data, device information, and metadata. The flexible data schema allows you to modify application data models without having to stick to a particular schema.

- You can also quickly query data by using a clustered index. A clustered index defines the order in which data is physically stored in a table and is great when it comes to performance. But other than clustered indexes in Azure Table Storage, it's possible to perform **Open Data-based queries** (**Odata-based queries**), and by using these queries, you can define the query options directly from the URL path.

- Working with Azure Table Storage is easy because you can either manipulate your data from Azure Storage Explorer or directly from a .NET or .NET Core application by using the Azure Tables client library for .NET. You can check it out at this link: `https://github.com/Azure/azure-sdk-for-net/blob/main/sdk/tables/Azure.Data.Tables/README.md`.

In the following diagram, we can see a key, which is unique and used to identify an item stored, and a value, which is a set of fields:

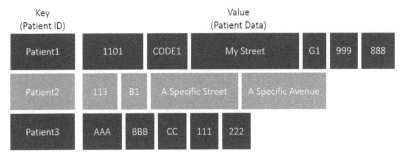

Figure 8.6 – Key value in Azure Table Storage

How does it work inside Azure Table Storage? It splits the table into a set of partitions, which facilitates rapid access to data. This mechanism groups the related rows that have a common property or partition key. Rows sharing the same partition key will be stored in the same place.

Inside these Azure Table Storage accounts, we can have tables. A table always represents either a collection of rows or a single row. There is no requirement to define the schema for a specific table, meaning you can store data with different properties within the same table. So, we said that inside storage accounts, we have tables. Now, inside tables, we have rows, and a row is a set of data with a single property or multiple properties. A row is similar to a database role in SQL databases.

A partition can be updated by adding or removing rows. A table includes any number of partitions.

The key in an Azure Table Storage table contains two elements: the key and the row key. The key is considered identified, and the row key is unique in the same partition.

Let's create a new Azure Table Storage table.

Creating a table in Azure Table Storage in the Azure portal

We have created an Azure Storage account before. It's time to learn how to create an Azure table.

Azure tables are used to store rows where we store the data. We will learn two different ways of creating Azure tables:

- The first one is manually using the Azure portal
- The second way is by using the ASP.NET Core application

To create an Azure Storage table, follow these steps:

1. Open your Azure Storage account and select **Storage browser**, and then click on **Tables**, as presented in the following screenshot:

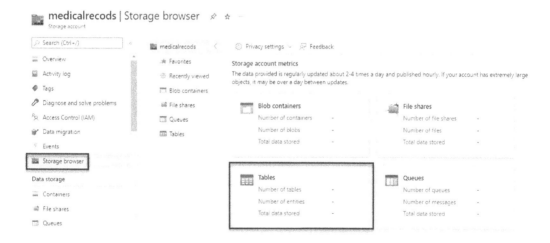

Figure 8.7 – Azure Storage account – Tables

2. Click on **Add table**. A dialog window will be displayed and we will add the table name:

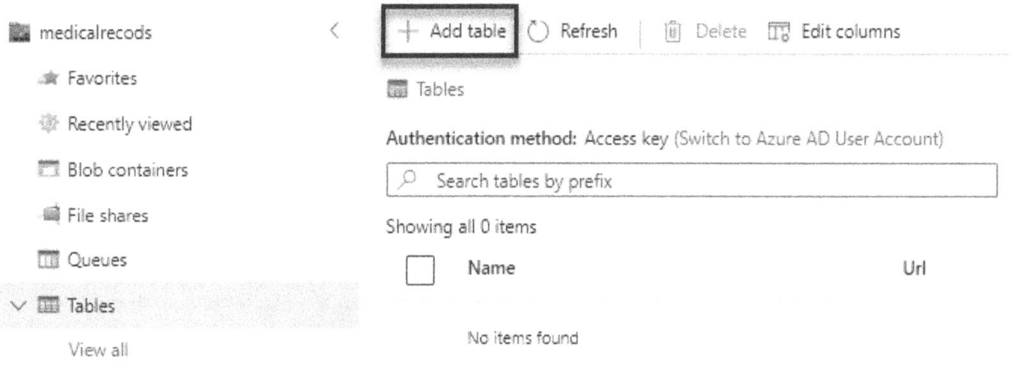

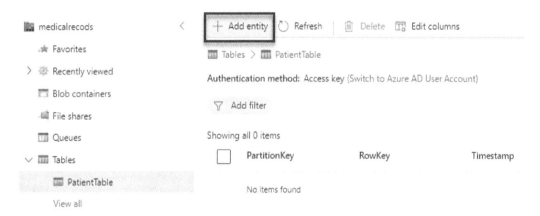

Figure 8.8 – Add table in an Azure Storage account

3. Now, we will select the table we created and add a new row, which in the Azure portal is called an entity. Click on **Add entity**:

Figure 8.9 – Add entity in an Azure Storage table

4. A new dialog window will be displayed to add **PartitionKey** and **RowKey**. We can also click on **Add property** to add a different property if needed:

- The **PartitionKey** property contains a string value that identifies the partition to which the entity belongs. Partitions are a vital part of table scalability. Entities with the same **PartitionKey** value are stored in the same partition.

- The **RowKey** property stores a string value that uniquely identifies the entity within each partition. **PartitionKey** and **RowKey** form the entity's primary key.

5.   Then, click on the **Insert** button:

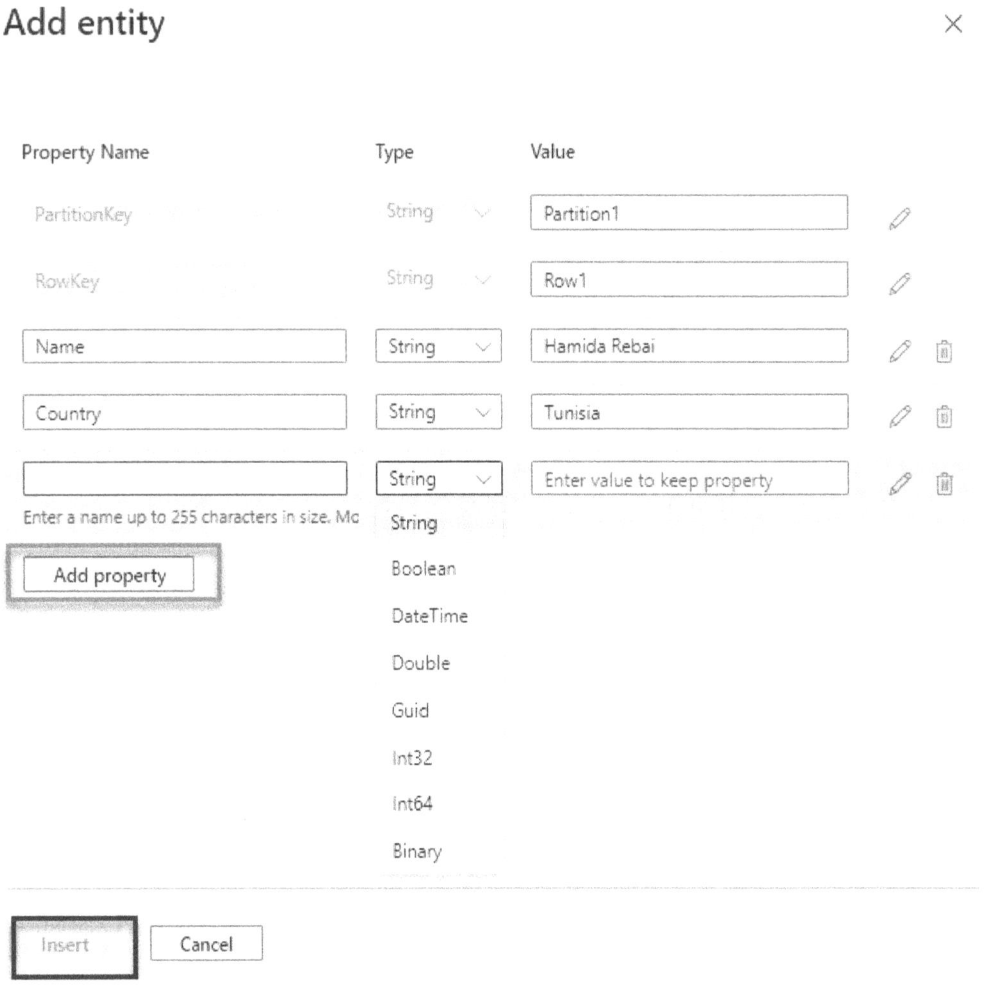

Add entity   ×

Property Name	Type		Value		
PartitionKey	String	∨	Partition1	✎	
RowKey	String	∨	Row1	✎	
Name	String	∨	Hamida Rebai	✎	🗑
Country	String	∨	Tunisia	✎	🗑
	String	∨	Enter value to keep property	✎	🗑

Enter a name up to 255 characters in size. Mc   String

Add property   Boolean

                                                   DateTime

                                                   Double

                                                   Guid

                                                   Int32

                                                   Int64

                                                   Binary

Insert   Cancel

Figure 8.10 – Add PartitionKey, RowKey, and properties

Check whether a new entity has been created after clicking **Insert**. The entity should contain the specified values, including a timestamp that contains the date and time that the entity was created:

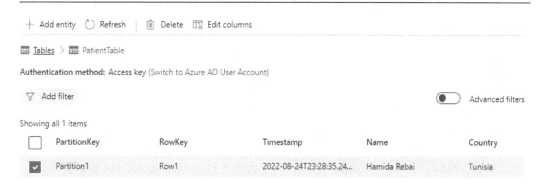

Figure 8.11 – Display an entity in an Azure Storage table

We can create this structure with different elements in our Azure Storage account:

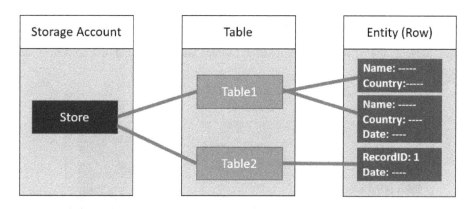

Figure 8.12 – Azure Table Storage components

We created an Azure Storage account in Azure. We added a new table, which is a collection of entities (rows). For every table, we can create a new entity that includes a set of properties. We added a partition key and a row key. We can also add a property that is presented as a name-value pair. We covered the different components of Azure Table Storage, so now we will move on to exploring Azure Blob Storage.

Exploring Azure Blob Storage

Azure Blob Storage provides massively scalable storage for unstructured data. Unstructured data is any data that isn't stored in the structured database format. This can be anything: binary, text files, images, or videos. Microsoft Azure Virtual Machines store the disk images in Blob Storage.

Blob is an acronym for **binary large object**.

Azure currently supports three different types of blobs:

- **Block blobs**: A block blob is managed as a set of blocks. The size of each block varies up to 100 MB. A block blob can contain up to 50,000 blocks. The maximum size is over 4.7 TB. A block is the smallest amount of data that can be read or written as a unit. Block blobs are great for storing separate large binary objects that rarely change.

- **Page blobs**: A page blob is organized as a collection of fixed-size 512-byte pages. A page blob is optimized to support random read and write operations; you can fetch and store data for a single page if necessary. A page blob can hold up to 8 TB of data. Azure uses page blobs to implement virtual disk storage for virtual machines.

- **Append blobs**: Append blobs are like block blobs but more optimized to support a new feature, that is, to append operations. You can only add blocks to the end of an append blob. Updating or deleting existing blocks is not supported. Each block can be of various sizes up to 4 MB. The maximum size of an append blob is slightly over 195 GB.

Blobs are stored in containers, which allows you to organize your blobs depending on your business needs.

This diagram presents the structure of Azure Blob Storage and how we can use an Azure account with containers and blobs:

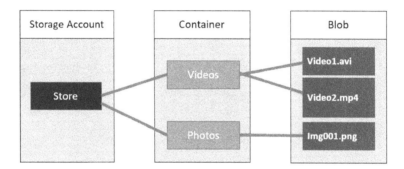

Figure 8.13 – Storage account – containers and blobs

Now, let's configure Blob Storage. We will use the same Azure Storage account:

1. Create a new container. Go to **Storage browser**, select **Blob containers**, and then click on **Add container**:

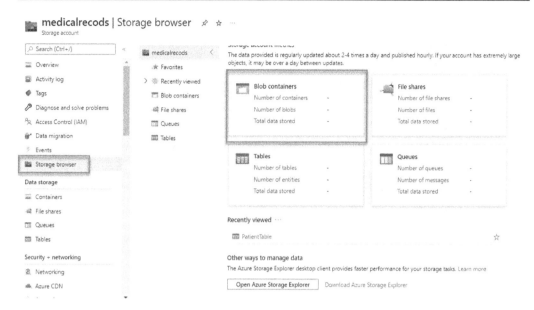

Figure 8.14 – Azure Storage account – Blob containers

2. Now, you can provide a name for your container and then select the public access level. You can select **Private**, **Blob**, or **Container**:

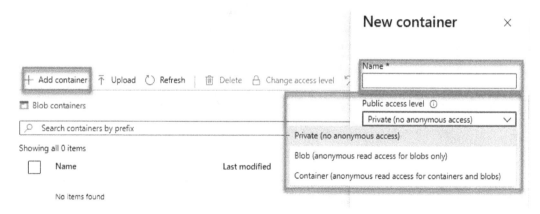

Figure 8.15 – Add a new container in Azure Blob Storage

3. You can add encryption if you select **Public access level** in the **Advanced** section. After that, click on the **Create** button.

A container is now created. When we select a container, we can manipulate it. We will open the container we just created and upload some files by selecting the **Upload** button. Select your file or files:

Figure 8.16 – Upload files in a container

To customize the file or files you upload, you can click **Advanced** and then select the authentication type, blob type, blob size, tier, and folder. We're going to leave the defaults. Once the file or files have been uploaded, you can then change the specific settings for that file.

You could view or edit the file, download it, see the properties, generate the shared access signature, view the snapshots, create snapshots, and change the tier if you wanted to do so. If you require a lock on the blob for read and write operations, then you can lease the blob by selecting **Acquire lease**. You can also easily delete the file from here:

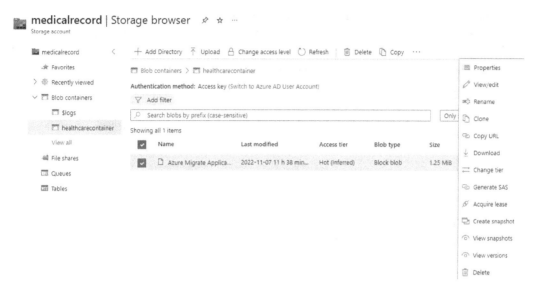

Figure 8.17 – Uploaded file settings

Let's go back and look at some of the other options that can be used with Blob Storage. You can map a custom domain name to your storage account. To do so, create the **Canonical Name** (**CNAME**) with your favorite registrar, then register the name with Azure and add it.

There are two options for doing this:

- The first option consists of setting up a custom domain.

- The second option consists of using a subdomain. There'll be no downtime when your storage account moves over to the custom domain name.

We talked about soft delete when we created the storage account, but we can also configure it at the blob level. You do so by selecting **Data protection**. Here, you can enable soft delete and then specify the number of days that the blob will be retained if it is deleted. To utilize the Azure **content delivery network** (**CDN**), you create a new CDN profile under Azure CDN. However, this is beyond the scope of this chapter.

Another feature of Azure Storage is the ability to use Azure Cognitive Search to index documents. You would set up this feature in Azure Search, just like CDN.

Finally, we have **Lifecycle management**. It's here that we can manage data based on rules.

On the Azure Blob Storage account home page, select **Lifecycle management** under **Data management**, and you'll see the screen shown in the following screenshot, where you select **Add a rule**:

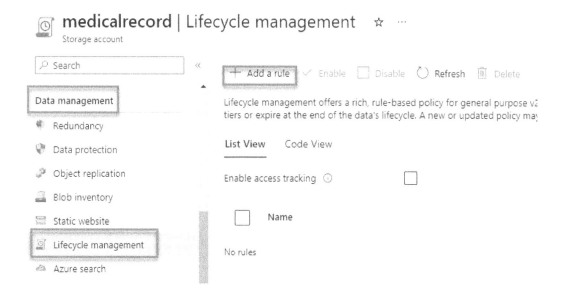

Figure 8.18 – Add a rule for Lifecycle management in an Azure Storage account

In the following figure, we illustrate life cycle management in an Azure Storage account:

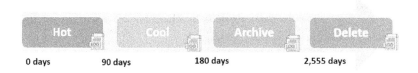

Figure 8.19 – Azure Blob Storage life cycle management

For example, we could create a rule that would move blobs to another access tier based on the number of days since they were last modified. We're going to come up with a rule name of `RuleTest`, as shown in the following screenshot:

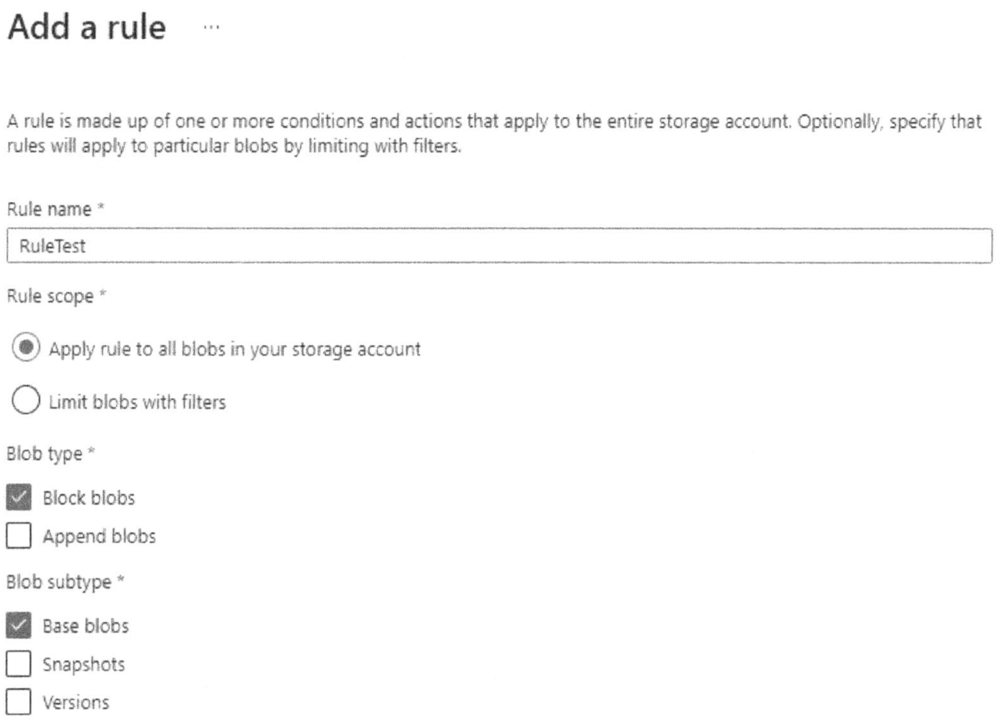

Add a rule   ⋯

A rule is made up of one or more conditions and actions that apply to the entire storage account. Optionally, specify that rules will apply to particular blobs by limiting with filters.

Rule name *

> RuleTest

Rule scope *

(●) Apply rule to all blobs in your storage account

(○) Limit blobs with filters

Blob type *

[✓] Block blobs

[☐] Append blobs

Blob subtype *

[✓] Base blobs

[☐] Snapshots

[☐] Versions

Figure 8.20 – Add a rule in an Azure Storage account

In the **Base blobs** tab, select the first option and move the blob to cool storage after, let's say, 60 days since it was last modified:

Figure 8.21 – Add a rule in an Azure Storage account – the Base blobs tab

We select **Add conditions** to enable moving data to archive storage after it hasn't been modified for a certain amount of time. Let's update it to 90 days and then delete it. In the following figure, we present the different rules added:

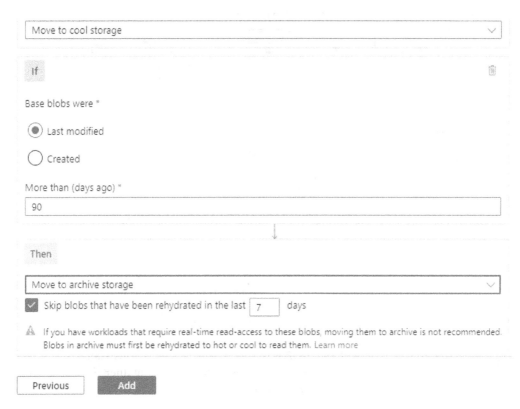

Figure 8.22 – Add new condition in Azure Storage account – the Base blobs tab

We have the option to create a filter set, which will limit the rules to certain objects. Finally, we select **Review and Add**. Now, that blob will be moved automatically through that life cycle based on the number of days since it was modified.

Azure Blob Storage client library for .NET

Blob storage is optimized to store a massive amount of unstructured data, and in this section, we will use a client library for .NET: Azure Blob Storage.

Follow these steps to install the package and work out example code for basic tasks:

1. Create a .NET console app using Visual Studio 2022 or the .NET CLI.

2. In order to interact with Azure Blob Storage, we need to install the Azure Blob Storage client library for .NET using NuGet Package Manager or Package Manager Console, as presented in the following screenshot:

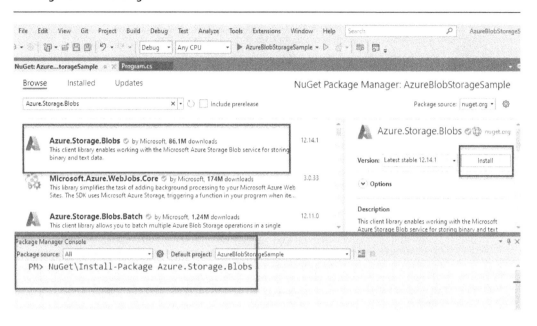

Figure 8.23 – Azure Blob Storage .NET library

3. In `Program.cs`, we will use the `BlobServiceClient` constructor using the account's connection string. To create a unique name for the container, we will use the `CreateBlobContainerAsync` method:

```
var blobServiceClient = new BlobServiceClient("<your-
storage-account-connection-string>");
BlobContainerClient containerClient = await
blobServiceClient.CreateBlobContainerAsync(containerName);
```

4. We will create a local directory for uploading and downloading files. We will write text in a file and get a reference to a blob using the `GetBlobClient` method:

```
string localPath = "datarepo";
Directory.CreateDirectory(localPath);
string fileName = "file-" + Guid.NewGuid().ToString() +
".txt";
string localFilePath = Path.Combine(localPath, fileName);
BlobClient blobClient = containerClient.
GetBlobClient(fileName);
```

5. We will upload data from the local file using the `UploadAsync` method:

```
await blobClient.UploadAsync(localFilePath, true);
```

6. To list all blobs in the container, we will use the `GetBlobsAsync` method:

```
await foreach (BlobItem blobItem in containerClient.
GetBlobsAsync())
{
    Console.WriteLine(«\t» + blobItem.Name);
}
```

7. To download the blob's contents and save it to a file, we will use the `DownloadToAsync` method:

```
await blobClient.DownloadToAsync(downloadFilePath);
```

8. If we need to delete the bloc container, we call the `containerClient` and the `DeleteAsync` method, and if we want to delete the local source and downloaded file, we call the `Delete` method for every file:

```
await containerClient.DeleteAsync();
File.Delete(localFilePath);
File.Delete(downloadFilePath);
```

If you need more details about this library, you can follow this link, `https://github.com/Azure/azure-sdk-for-net/tree/main/sdk/storage/Azure.Storage.Blobs/samples`, which includes the different samples to be tested for Azure Blob Storage.

We've discussed Azure Blob Storage, and in the next section, we'll move on to the basic concepts of Azure Disk Storage.

Exploring Azure Disk Storage

Azure Disk Storage provides managed storage for virtual machine disks. But applications and services are allowed to use these disks as needed. In Disk Storage, data is persistently stored and accessed from an attached virtual hard disk.

We have different disk sizes and performance levels, from **solid-state drives** (**SSDs**) to traditional **hard disk drives** (**HDDs**).

Azure Managed Disks uses redundancy to achieve availability. The user is allowed to create up to 50,000 virtual machine disks within a region. Consider your capacity and performance requirements when using Azure Managed Disks. The cost depends on the type of storage hardware and the size of the virtual disk.

We've defined Azure Disk Storage, so now we'll move on to Azure Files in the next section.

Exploring Azure Files

Azure Files storage provides cloud-based file shares, which can be accessed from the cloud or on-premises.

Azure Files storage offers shared storage for applications using the industry-standard **Server Message Block (SMB)** protocol and the **Network File System (NFS)** protocol. Azure Virtual Machines and cloud services can share file data across application components via mounted shares. On-premises applications can also access file data under sharing repositories. That means multiple virtual machines can share the same files with both read and write access. You can also read the files using the **REpresentational State Transfer (REST)** API interface or the storage client libraries. Applications running in Azure Virtual Machines, Windows, Linux, or other cloud services can mount a file storage share to access file data, just as an application would mount a typical SMB or NFS share. Any number of Azure Virtual Machines or roles can mount and access the file storage share simultaneously. Azure Files SMB file shares are accessible from Windows, Linux, and macOS clients. NFS file shares are accessible from Linux and macOS clients.

The common uses of file storage

Azure Files can be used to completely replace traditional on-premises file servers or **Network Attached Storage (NAS)** devices. Popular operating systems, such as Windows, macOS, and Linux, can directly mount Azure file shares wherever they are in any location. Azure Files makes it easy to lift and shift applications to the cloud that expect a file share to store file applications or user data. Azure file shares can also be replicated with Azure File Sync to Windows servers, either on-premises or on the cloud. To ensure performance and distributed caching of data where it is used, shared application settings are stored, for example, in configuration files, and diagnostic data such as logs, metrics, and crash dumps to a shared location. We store the tools and utilities needed to develop or administer Azure virtual machines or cloud services. That has covered all the common use cases of file storage for Azure Files.

Adding Azure Files in the Azure portal

Now, we will see how we add Azure Files in the Azure portal:

1.   Go back to **Storage browser** and select **File shares**:

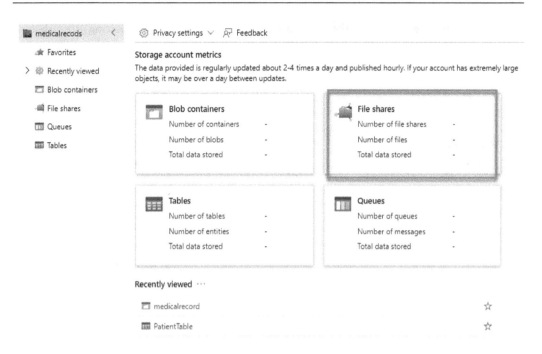

Figure 8.24 – Storage browser | File shares

2. Select **Add file share** and add a name. Select **Tier**, then **Transaction optimized** or **Hot** or **Cool**:

    - **Transaction optimized** is used to enable transaction-heavy workloads. We use it for applications requiring file storage as a backend store.

    - **Hot** is the optimized option for some file-sharing scenarios; for example, if a team is sharing files, Azure Files will synchronize them.

    - **Cool** is used for archive storage scenarios.

Once created, we can explore the different settings, as presented in the following screenshot:

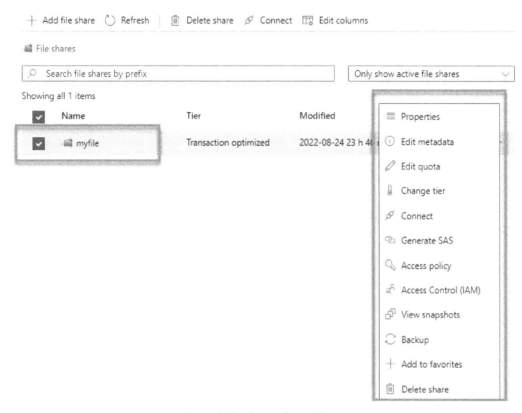

Figure 8.25 – Azure Files settings

3. If we select **Connect**, we can configure the connection to the Azure file share from Windows, Linux, or macOS. To secure the connection, select **Active Directory** or **Storage account key** for Windows. We can change **Tier** from **Transaction optimized** to **Cold**, for example, or from **Cold** to **Hot**.

Sometimes, it's not easy to decide when to use Azure file shares instead of blobs or disk shares in your solution. If we compare the different features, we can decide which solution is suitable. If we compare Azure Files to Azure Blob side by side, we see that Azure Files provides an SMB and NFS interface, client libraries, and the REST interface, which allows access from anywhere to store files. Azure Blob provides client libraries and the REST interface, which allows unstructured data with flat namespaces to be stored and accessed at a massive scale in the blobs.

When to use Azure files versus blobs

You can use Azure files to lift and shift an application to the cloud, which already uses the native filesystem APIs to share data between it and other applications running in Azure. You want to store development and debugging tools that need to be accessed from many virtual machines. Azure blobs, on the other hand, can be used if you want your application to support streaming and random-access scenarios and you want to be able to access application data from anywhere. There are a few other distinguishing features on when to select Azure files over Azure blobs. Azure files are true directory objects, while Azure blobs are a flat namespace. Azure files can be accessed through file shares. However, Azure blobs are accessed through a container. Azure files provide shared access across multiple virtual machines, while Azure disks are exclusive to a single virtual machine.

When you select which storage feature to use, you should also consider pricing and perform a cost-benefit analysis.

Azure Storage is accessible via a REST API or designed for Microsoft Azure Virtual Machines, as illustrated in the following diagram:

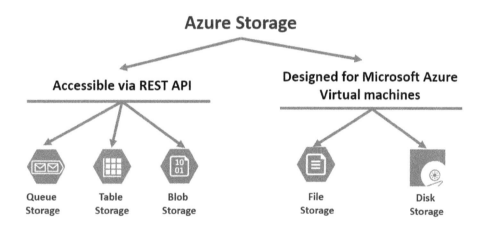

Figure 8.26 – Azure Storage

The Microsoft Azure Storage services REST API provides programmatic access to blob, queue, table, and file services in your Azure or development environment through a storage emulator. You can see more details about these APIs at this link: `https://learn.microsoft.com/en-us/rest/api/storageservices/`.

Summary

In this chapter, we presented the different storage options that are available in Azure Storage services. We started by creating an Azure Storage account, which includes different types of storage. We explored Azure Table Storage, Azure Blob Storage, Azure Disk Storage, and, in the end, Azure Files.

In the next chapter, we will cover designing and implementing cloud-native applications using Microsoft Azure Cosmos DB.

Further reading

For more information, you can check out the Microsoft documentation at this link: `https://learn.microsoft.com/en-us/azure/storage/`. If you need to follow a example using Azure Storage with .NET library, use this link: `https://learn.microsoft.com/en-us/azure/storage/common/storage-samples-dotnet`.

Questions

1. What are the common uses of file storage?

2. When do you use Azure files versus blobs?

Working with Azure Cosmos DB to Manage Database Services

We store data in relational tables if we are using relational databases. Unstructured data can be stored in a storage solution, but sometimes, the structure of a relational database can be too rigid and generally leads to poor performance in certain cases, requiring the implementation of specific and detailed tuning. There are several models, known as NoSQL databases, that present solutions for certain scenarios, such as documents, graphs, column family stores, and key-value stores.

NoSQL databases are defined by multiple characteristics: they are non-relational, have a **JavaScript Object Notation (JSON)** schema, and are designed for scaling out.

This chapter will cover designing and implementing cloud-native applications using a multi-model NoSQL database management system, which is **Microsoft Azure Cosmos DB**.

In this chapter, we're going to cover the following main topics:

- NoSQL databases
- Exercise 1 – creating an Azure Cosmos DB account using the Azure portal
- Exploring the Azure Cosmos DB SQL API
- Exercise 2 – creating an Azure Cosmos DB SQL API account
- Exercise 3 – connecting to the Azure Cosmos DB SQL API with the SDK

NoSQL databases

Organizations now handle more than structured data. NoSQL databases store unstructured data in documents rather than relational tables, where we store structured data. So, we categorize them as "more than just SQL" and decompose them into various flexible data models. NoSQL database types include document-only databases, key-value stores, wide-column databases, and chart databases.

NoSQL databases are built from the ground up to store and process vast amounts of data at scale, supporting the ever-growing modern business.

The following are definitions for the most popular types of NoSQL databases:

- **Key-value store**: A key-value store groups related data into collections so that records can be identified by unique keys for easy retrieval. A key-value store retains the benefits of a NoSQL database structure, yet has enough structure to reflect the values of relational databases (as opposed to non-relational databases). We can see an example in the following figure:

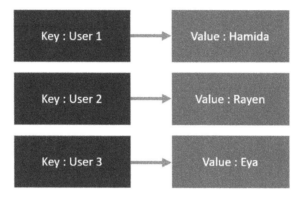

Figure 9.1 – Key-value NoSQL database type

Microsoft Azure Table Storage is an example of this type of NoSQL database.

- **Document**: Document-only databases are primarily created to store document-centric data as documents, such as JSON documents. These systems can also be used to store XML documents such as NoSQL databases. Azure Cosmos DB is an example of this type.

- **Table style**: This type is a data model based on sparsely populated tables. They are also called wide-column databases. They use the tabular format of relational databases, but the naming and formatting of the data in each row can vary greatly, even in the same table. Similar to key-value stores, wide-column databases have a basic NoSQL structure while retaining a high degree of flexibility.

- **Graph**: The graph type is a record generally stored in key-data format. We use it to define the relationships between stored data points. Graph databases allow us to identify the patterns in unstructured and semi-structured information. In the following figure, we can see the different graph components:

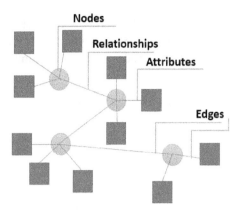

Figure 9.2 – Graph NoSQL database type

Polyglot persistence across multiple non-relational (NoSQL) data stores can add significant complexity and overhead. As a result, Microsoft provides a multi-model, geographically distributed database called Azure Cosmos DB with data structures for natively storing key-value pairs, JSON hierarchies, and graphs.

Azure Cosmos DB is a fully managed PaaS offering and its internal storage is abstracted from the different users.

Exercise 1 – creating an Azure Cosmos DB account using the Azure portal

Azure Cosmos DB is a distributed database engine with a core set of features no matter how you use your database. These features include the ability to elastically distribute databases, the ability to scale both storage and throughput bi-directionally, low latency with financially backed SLAs, various consistency options, and enterprise-class security features.

Azure Cosmos DB is designed for high responsiveness and always-on availability for most modern applications in the cloud. Cosmos DB is considered a NoSQL database and works primarily with four data models (key-value, documents, column-family or table-style, and graph).

Azure Cosmos DB includes documents stored where data is schemeless and most likely stored as JSON documents. Many document databases use JSON to represent the document structure, graph-oriented models where data is represented as diagrammatic structures such as nodes and edges, and key-value stores in their simplest form. A database management system stores only key-value pairs and has a wide column store that can store data with many dynamic columns. Cosmos DB's focus on high availability, geographic distribution, and speed provides some pretty cool advantages.

Cosmos DB manages data as a partitioned set of documents. A document is made up of a set of fields, and each field is identified by a key. The fields in each document may vary and a field may contain child documents. In a Cosmos DB database, the documents are organized into containers that are grouped into partitions.

A document can contain up to 2 MB of data, including small binary objects. But if you need to store larger blobs related to a document, you'll need to use Azure Blob Storage. You can then add a reference to the blob created in the document.

Let's start by creating an Azure Cosmos DB resource by using the Azure portal. We will follow these steps:

1. On the home page of the Azure portal, select **Create a resource**. You can use the search input and enter Cosmos or you can select the **Databases** category. Select **Create** under **Azure Cosmos DB**:

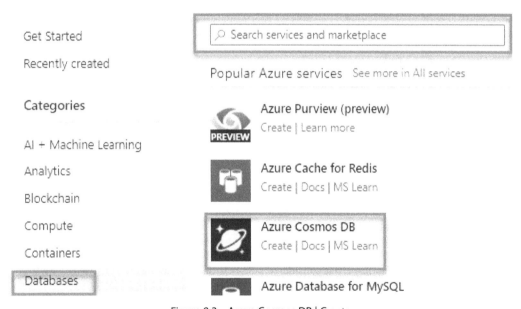

Figure 9.3 – Azure Cosmos DB | Create

2. On the **Select API option** page, we have to select an API to create a new account. Note that we are not able to change it after account creation:

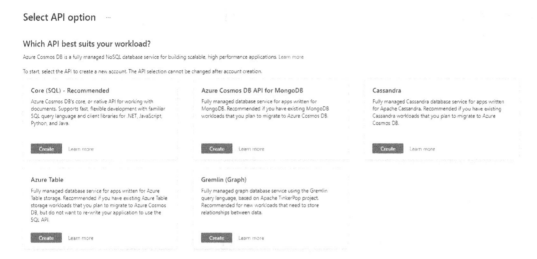

Figure 9.4 – The Select API option page in Cosmos DB

We have these APIs to choose from:

- **Core (SQL)** is recommended for new applications. It is a native API used for documents. It's Azure Cosmos DB's core.

- **Azure Cosmos DB API for MongoDB** is a native SQL API with APIs for MongoDB. MongoDB is a document database. This API is recommended if you already have an existing MongoDB workload and you would like to migrate it to Azure, for example.

- **Cassandra** is a fully managed database service dedicated to Apache Cassandra apps. If you already have an existing Cassandra workload, you can use this API to migrate it to Azure. Cassandra is a distributed NoSQL database management system used for large data.

- **Azure Table** is also a fully managed database service for Azure Table Storage apps. If we decide to migrate an existing Azure Table Storage resource to Cosmos DB, we can use this API.

- **Gremlin (Graph)** is Apache TinkerPop's graph traversal language. Gremlin is a functional dataflow language that allows users to concisely express complex traversals (or queries) of an application's property graph. Cosmos DB is a fully managed graph database service that uses Gremlin. If we would like to add new workloads that need to store relationships between data, this option is recommended.

3. Select any API option. We will have the same options on the **Create Azure Cosmos DB Account - Core (SQL)** page. Start with the **Basics** tab:

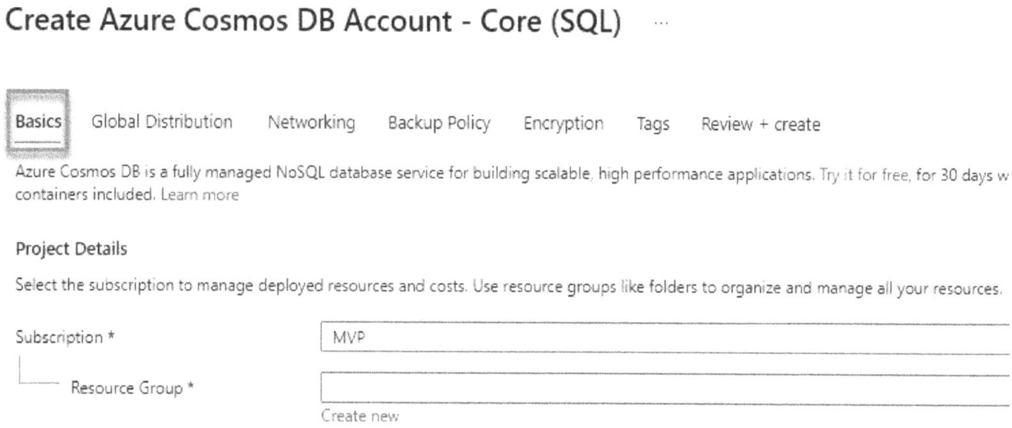

Create Azure Cosmos DB Account - Core (SQL)   ...

| Basics | Global Distribution | Networking | Backup Policy | Encryption | Tags | Review + create |

Azure Cosmos DB is a fully managed NoSQL database service for building scalable, high performance applications. Try it for free, for 30 days w containers included. Learn more

Project Details

Select the subscription to manage deployed resources and costs. Use resource groups like folders to organize and manage all your resources.

Subscription *   MVP

└─── Resource Group *
                           Create new

Figure 9.5 – Project Details on the Basics tab

On the **Basics** tab, we have the project details, which include common general information related to any Azure services, such as **Subscription** and selecting a resource group. Then, we can add the information needed for the instance details, such as the Cosmos DB account name, the location, and the capacity mode for your database operations. We have two different capacity modes to choose from:

- **Provisioned throughput**: You can configure the amount of throughput you expect the database to deliver, but you are required to plan the best performance level for the application's needs. This mode is expressed in **request units per second** (**RU/s**). There are two types: *standard (manual)* and *autoscale*. An Azure Cosmos DB SQL API database is a schema-agnostic set of container management units. Each container is a scalable unit of throughput and storage. Throughput can be provisioned at the database level, container level, or both. Throughput provisioned to a database is shared by all containers within the database. All containers share throughput resources, so we may not get predictable performance for certain containers in our database:

Cosmos DB Account
Database
Database
Container
Container
Container

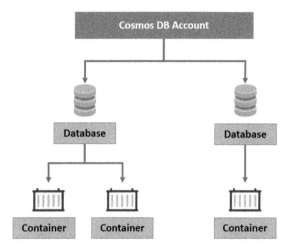

Figure 9.6 – Azure Cosmos DB components

- **Serverless**: We select this option to create an account in serverless mode, which means that we can run our database operations in containers without the need to configure previously provisioned capacity. It is a consumption-based model where each request consumes request units. The consumption model eliminates the need to pre-provision throughput request units:

> **Important note**
>
> When we are working with Azure Cosmos DB, we typically express database options in terms of cost expressed in RU/s.

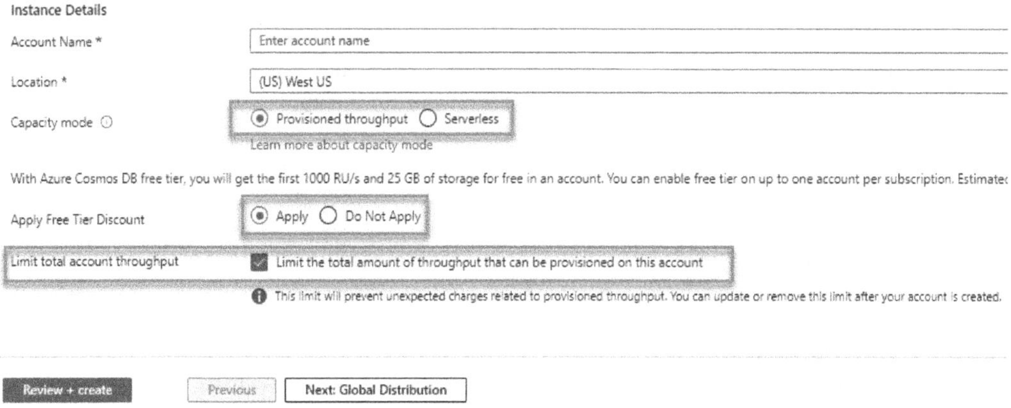

Instance Details
Account Name *
Enter account name
Location *
(US) West US
Capacity mode
Provisioned throughput
Serverless
Learn more about capacity mode
With Azure Cosmos DB free tier, you will get the first 1000 RU/s and 25 GB of storage for free in an account. You can enable free tier on up to one account per subscription. Estimate
Apply Free Tier Discount
Apply
Do Not Apply
Limit total account throughput
Limit the total amount of throughput that can be provisioned on this account
This limit will prevent unexpected charges related to provisioned throughput. You can update or remove this limit after your account is created.
Review + create
Previous
Next: Global Distribution

Figure 9.7 – Instance Details on the Basics tab

4. We will select whether we want to apply the Azure Cosmos DB free tier discount. We will get the first 1,000 RU/s and 25 GB of storage for free after creating this account.

    We can only create one free tier Azure Cosmos DB account per Azure subscription. If we are already using it, this option will not be displayed when we create a new Cosmos DB account.

5. It is recommended to select **Limit the total amount of throughput that can be provisioned on this account** to prevent any unexpected charges related to provisioned throughput.

6. In the **Global Distribution** tab, we can enable geo-redundancy and multi-region writes:

    A. Geo-redundancy is the ability to add more regions to your account

    B. The **Georedundancy writes** capability allows you to use your provisioned throughput for databases and containers around the world

7. On the **Networking** tab, select a connectivity method to connect the Cosmos DB account to a public endpoint via public IP addresses, a private endpoint, or to all networks.

8. We also have the **Backup Policy** tab, **Encryption** tab, and **Tags** tab. You can update them according to your needs or keep their defaults as they are.

9. After selecting **Create**, select **Review + create**.

Once the deployment has been completed, you can check the Azure Cosmos DB account page and start exploring the different elements to use it.

With that, we have created an Azure Cosmos DB account by using the Azure portal. Azure Cosmos DB includes multiple APIs. In the next section, we will start exploring the Cosmos DB SQL API.

Exploring the Cosmos DB SQL API

Azure Cosmos DB supports multiple APIs models, which makes it possible for you to select the ideal API for your application. In this section, we will specifically drill down into this SQL API. The Cosmos DB SQL API is the default Azure Cosmos DB API. You can store data using JSON documents to represent your data. Even though we are using the SQL API, we will still get to take advantage of all of the core features of Cosmos DB that are universal across all APIs.

The Core API for Cosmos DB is recommended when you're building new solutions. This API uses SQL as a query language, along with JavaScript as its programming language. Both of these languages are pretty universal and popular, making it more likely that you already have experience with both.

Talking about the SQL query language specifically, you can write queries against Cosmos DB using the same SQL syntax you would use with products such as Microsoft SQL Server. Even though you are using the SQL query language, you can use these queries with your JSON documents, so you don't have to do anything special to make your queries work with JSON. You can embed JSON objects directly into your queries. If you take a look at some unusual scenarios, you need to shape your JSON documents and test specific schema for the query result.

JavaScript is the language used by Cosmos DB for server-side programming. Storage procedures, triggers, and user functions are all written using JavaScript. Since JavaScript can natively process JSON objects, you'll find that JavaScript is ideal for manipulating properties and documents throughout your server-side code.

Exercise 2 – creating an Azure Cosmos DB SQL API account

We will create an Azure Cosmos DB account by following the same steps in the first exercise, but we will select the **Core (SQL)** API option.

We will start by adding a new database and a container. After, we will add data to the created database. We will query the data and, in the end, use the SDK of Cosmos DB to connect to the Azure Cosmos DB SQL API while using C# as the programming language.

Adding a new database and a new container

Let's look at the steps:

1. Select **Data Explorer** to create a new container, then select **New container**:

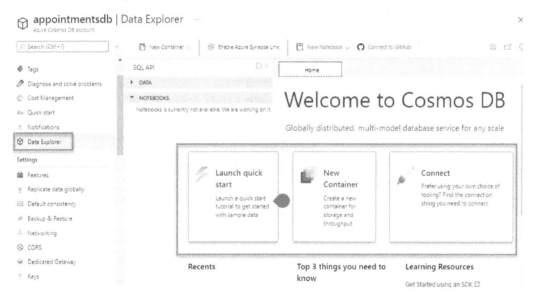

Figure 9.8 – Data Explorer option in the Azure Cosmos DB account

We need to fill in all the information needed:

- **Database id** is the name of the database; we can add any unique name or we can use an existing one.

- Check **Share throughput across containers**. This allows you to distribute your database's provisioned throughput across all containers in your database. We are using this option to save costs.

- Regarding **Database throughput (400 - unlimited RU/s)**, we can select **Autoscale** or **Manual**. We will leave the throughput at **400** required RU/s:

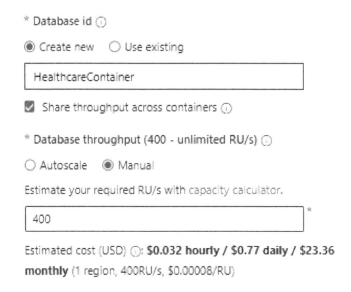

Figure 9.9 – Creating a new container – step 1

2. The second step is related to the container and the partition key. We will add the following information:

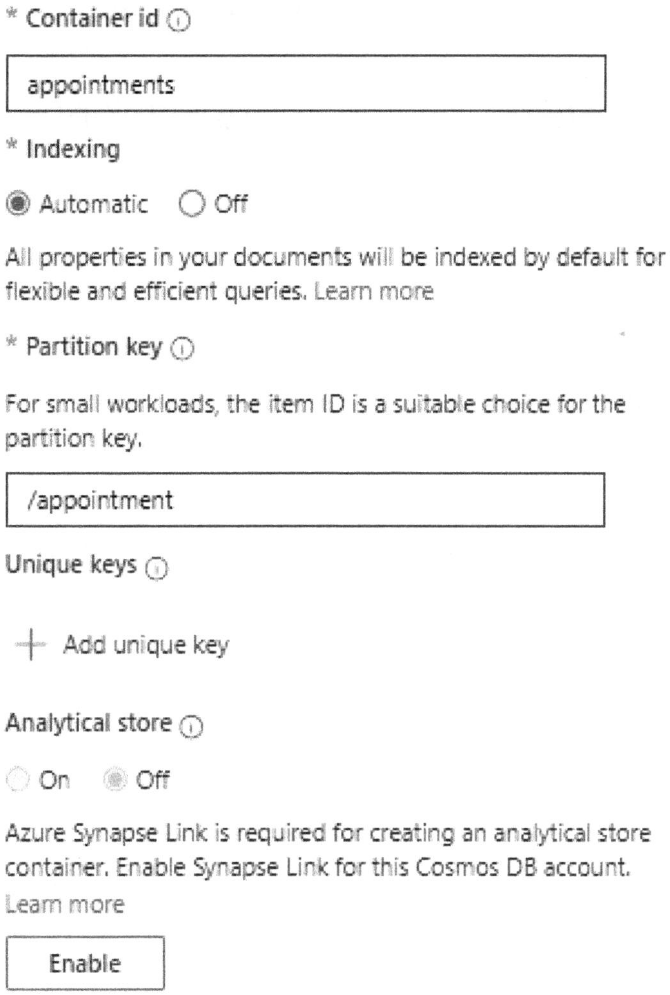

* Container id ⓘ

| appointments |

* Indexing

◉ Automatic   ○ Off

All properties in your documents will be indexed by default for flexible and efficient queries. Learn more

* Partition key ⓘ

For small workloads, the item ID is a suitable choice for the partition key.

| /appointment |

Unique keys ⓘ

╋ Add unique key

Analytical store ⓘ

○ On   ◉ Off

Azure Synapse Link is required for creating an analytical store container. Enable Synapse Link for this Cosmos DB account. Learn more

| Enable |

Figure 9.10 – Creating a new container – step 2

The container ID is the name of your container, and the partition key is used to automatically distribute data across partitions to ensure scalability.

3. Click on **OK** to confirm the creation of a new container. Once one has been created, **Data Explorer** will display it and the new database:

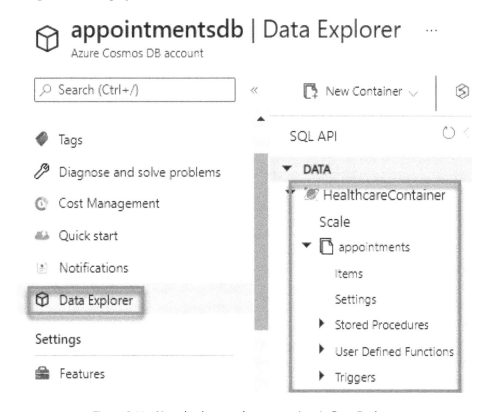

Figure 9.11 – New database and new container in Data Explorer

Now, we'll add data to our database. To do that, we will use JSON content; we will see how in the next section.

Adding data to a database

We will select **Items** under **appointments** and use the following structure to add data to the document:

```
{
    "id": "1",
    "appointment": "appointment with a dentist",
    "patientname": "Hamida Rebai",
    "doctorname": "Rayen Trabelsi",
```

```
    "isConfirmed": false,
    "date": "2023-03-09T22:18:26.625Z"
}
```

On the right of the **Documents** pane, add the JSON content and click on **Save**. You can add more than one item by clicking on **New Item**:

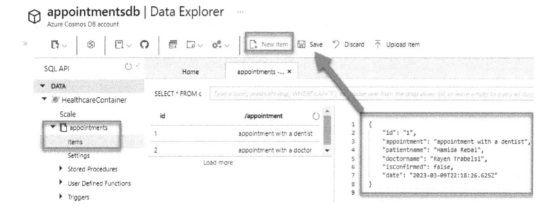

Figure 9.12 – Adding data to your database using JSON content

Note that we don't have a schema for the data added in Azure Cosmos DB.

We have added data to our database using JSON content. We are now able to query the data.

Querying data

We will use queries in **Data Explorer** to search for and filter data according to a specific criterion.

We will select **New SQL Query** at the top. We can add a default selection that will display all the data in our database:

```
SELECT * FROM c
```

In the next figure, we will select **New SQL Query** to add the SQL command and to execute it, we select **Execute Query**:

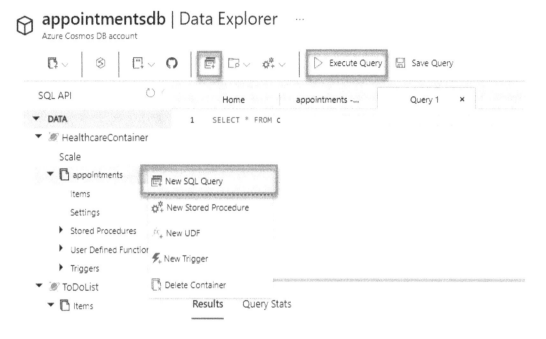

Figure 9.13 – Querying data from documents in Cosmos DB

We can add more filters to retrieve data. For example, if we need to order the result, we can use this query:

```
SELECT * FROM c ORDER BY c.id DESC
```

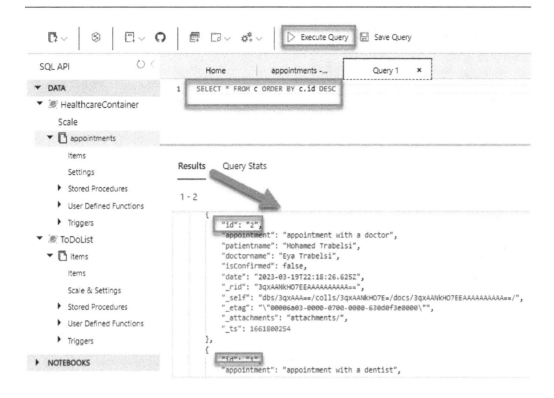

Figure 9.14 – Filtering data by order – DESC

You can use any SQL queries to search and retrieve data. You can also add stored procedures and triggers.

Data Explorer is an easy-to-use tool in the Azure portal that is used to add queries in SQL. We observe the results in JSON format, and we can measure the impact in terms of RU/s using the **Query Stats** tab.

We can generate metrics in a CSV file:

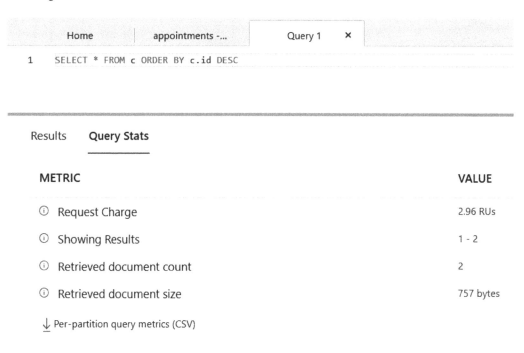

Figure 9.15 – Query Stats in Data Explorer

In this section, we created a new Azure Cosmos DB container. We used the SQL API, and we added a new database and a new container. We added data to the database to test Data Explorer and retrieved the data using the SQL language.

In the next section, we will learn about the SDK, which we will use to connect to the Azure Cosmos DB SQL API using C# as the programming language.

Exercise 3 – connecting to the Azure Cosmos DB SQL API with the SDK

In this section, we will explore Microsoft .NET SDL v3 for Azure Cosmos DB and explore the different methods and classes used to create different resources.

It is really simple to add the .NET SDK to your solution. Once we add the SDK to the project using NuGet by adding the **Microsoft.Azure.Cosmos** package, we will create a CosmosClient. We need an endpoint and a master key for a Cosmos DB account, which we will retrieve and add to the CosmosClient constructor. Now that the CosmosClient class has been configured, we simply call methods on it to do whatever we need. We can create databases and containers, query documents, and run stored procedures, which means that we can do all operations that we can do in Data Explorer.

But how does it work? The SDK figures out how to set up the HTTP headers, make the REST call, and parse the response back from Cosmos DB. There is great flexibility in the types of objects we can use to represent our documents. These can be typical C# classes or **Plain Old CLR Object** (**POCOs**) or plain old **Common Language Runtime** (**CLR**) objects, or we can use dynamic types with no predefined properties, which fits with the flexible schema that we have in a document database. Most operations performed will be asynchronous, where the code doesn't get blocked while waiting for a response from Cosmos DB after making a request. That's where the Task Parallel Library in .NET comes in.

> **Important note**
>
> If you prefer using Language-Integrated Query (LINQ) rather than SQL, the SDK also includes a LINQ provider that can translate LINQ queries that you write in C# into Cosmos DB SQL for execution.

Let's start by creating a new console application using .NET Core 6. This will be a simple demo that just runs a query and displays some results:

1. Grab the SDK from NuGet, search for `Microsoft.Azure.Cosmos`, and install it. Add the NuGet **Microsoft.Extensions.Configuration.Json** package so that we can use a JSON configuration file to store our Cosmos DB connection information:

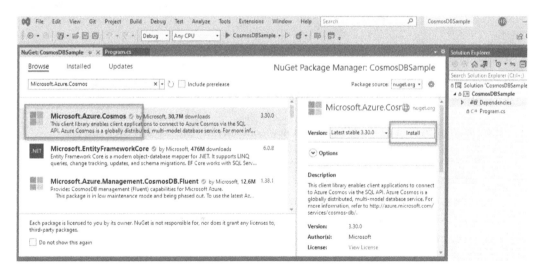

Figure 9.16 – Adding Microsoft.Azure.Cosmos

2. We will introduce the code snippet that will add the endpoint URI and the primary key. We will replace the information related to the endpoint URI and the primary key in `Program.cs`:

```
private static readonly string EndpointUri = "Add your
endpoint here";
```

```
private static readonly string PrimaryKey = "add your
primary key>";
```

This information can be found in **Keys** under **Settings** in the Azure Cosmos DB account. This is an example:

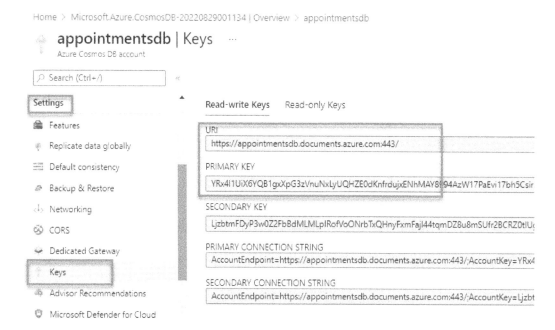

Figure 9.17 – Getting the primary URI and primary key from Keys under Settings

3. If we want to create a new database and a new container, we can use this code snippet:

```
private CosmosClient cosmosClient;
private Database database;
private Container container;
private string databaseId = "appointmentDB";
private string containerId = "appointmentContainer";

// Runs the CreateDatabaseAsync method
    await this.CreateDatabaseAsync();

    // Run the CreateContainerAsync method
    await this.CreateContainerAsync();
private async Task CreateDatabaseAsync()
```

```
{
    // Create a new database using the cosmosClient
    this.database = await this.cosmosClient.
CreateDatabaseIfNotExistsAsync(databaseId);
}
private async Task CreateContainerAsync()
{
    // Create a new container
    this.container = await this.database.
CreateContainerIfNotExistsAsync(containerId, "/
LastName");
}
```

We can add a new JSON file that will include the different settings. We will optimize our code to perform a query in the documents. So, we will create a new JSON file similar to the one shown in the following figure:

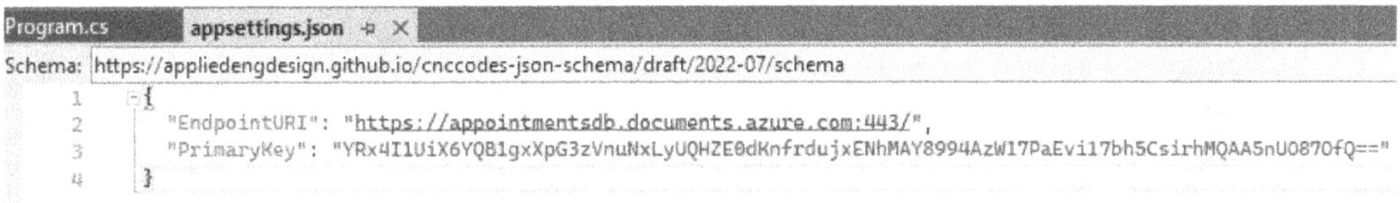

```
Program.cs        appsettings.json  -þ ×
Schema: https://appliedengdesign.github.io/cnccodes-json-schema/draft/2022-07/schema
    1    -{
    2        "EndpointURI": "https://appointmentsdb.documents.azure.com:443/",
    3        "PrimaryKey": "YRx4I1UiX6YQB1gxXpG3zVnuNxLyUQHZE0dKnfrdujxENhMAY8994AzW17PaEvi17bh5CsirhMQAA5nU0870fQ=="
    4    }
```

Figure 9.18 – App settings in a JSON file

The Program class will be similar to this:

```
using Microsoft.Azure.Cosmos;
using Microsoft.Extensions.Configuration;

var config = new ConfigurationBuilder().
AddJsonFile("appsettings.json").Build();
var endpointuri = config["EndpointURI"];
var primarykey = config["PrimaryKey"];

using (var client = new CosmosClient(endpointuri, primarykey))
{
    var container = client.GetContainer("HealthcareContainer",
"appointments");
    var mysqlquery = "SELECT * FROM c";
```

```
    var iterator = container.GetItemQueryIterator<dynamic
>(mysqlquery);
    var page = await iterator.ReadNextAsync();


    foreach (var item in page)
    {
        Console.WriteLine(item);
    }
}
```

We will debug the solution to display the different items. You can display the information that you need inside foreach:

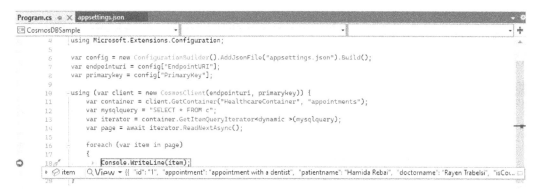

Figure 9.19 – Getting items using CosmosClient and a SELECT query

If we want to use LINQ instead of a SQL query, it is simple: you need to add a class that will include your object. In our case, I added the Appointment class; we will replace the code with the following code:

```
using (var client = new CosmosClient(endpointuri, primarykey))
{
    var container = client.GetContainer("HealthcareContainer",
"appointments");
    var q = from d in container.GetItemLinqQueryable
<Appointment>(allowSynchronousQueryExecution:true)
 select d;


    var documents = q.ToList();
    foreach(var document in documents)
```

```
    {
        var d = document as dynamic;
        Console.WriteLine(d);
    }
}
```

Azure Cosmos DB also provides language-integrated transactional execution of JavaScript, which allows you to write stored procedures or triggers, user-defined functions, and much more.

Azure Cosmos DB is a distributed database system designed for low latency, elastic throughput scalability, well-defined semantics for data consistency, and high availability. It enables data reading and data writing from a local replica of your database. Azure Cosmos DB replicates data across all regions associated with your Azure Cosmos DB account.

If we build an application that requires a fast response time anywhere in different regions, with unlimited elastic scalability of throughput and storage and high availability, we will use Azure Cosmos DB.

You have to configure your databases to be globally distributed and available in any Azure region to achieve high availability with Azure Cosmos DB. Azure Cosmos DB is a managed multi-tenant service. It manages all details of individual compute nodes transparently. So, users don't have to worry about patching and planned maintenance. We also have replica outages, which refer to a single node failure in an Azure Cosmos DB cluster deployed in an Azure region. Azure Cosmos DB automatically mitigates replica failures by guaranteeing at least three replicas of your data in each Azure region of your account within a quorum of four replicas.

In the next section, we will connect an application's ASP.NET Core Web API deployed on Azure App Service with Azure Cosmos DB as a database.

Exercise 3 – connecting Azure App Service with Azure Cosmos DB

In this section, we will connect Azure App Service with Azure Cosmos DB as a database. An ASP.NET Web API will connect to a Cosmos DB database to manage the appointments booked by patients before sending them to the doctor's dashboard.

We will use Visual Studio 2022 with the Azure workload installed. We will follow these steps to connect to Azure Cosmos DB using **Connected Services**:

1. In Solution Explorer, right-click the **Connected Services** node, and, from the context menu, select **Add Connected Service**:

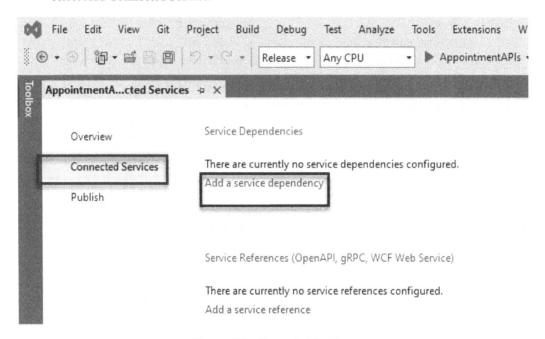

Figure 9.20 – Connected Services

2. In **Connected Services**, select **Add a service dependency** or the + icon for **Service Dependencies**.

3. On the **Add dependency** page, select **Azure Cosmos DB**:

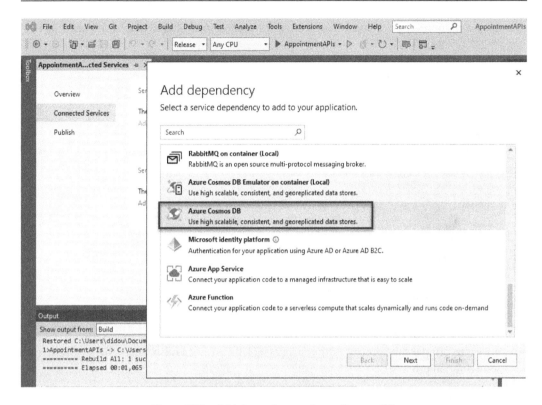

Figure 9.21 – Add dependency – Azure Cosmos DB

4.   Select **Next**. On the **Connect to Azure Cosmos DB** page, select an existing database, and select **Next**. We can create a new Azure Cosmos DB instance by selecting + **Create new**:

Figure 9.22 – Connect to Azure Cosmos DB

5. Enter a connection string name, and select a connection string stored in a local secrets file or Azure Key Vault:

Figure 9.23 – Configuring a connection string for Azure Cosmos DB

6. The **Summary of changes** screen displays all the modifications that will be made to the project to complete the process. If the changes look fine, select **Finish** to start the dependency configuration process:

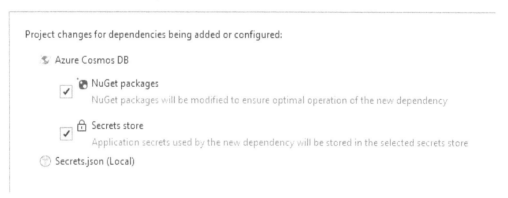

Figure 9.24 – Summary of changes in Visual Studio 2022

The result will be similar to what's shown in the following figure, where all service dependencies have a **Connected** status:

Figure 9.25 – Service Dependencies on the Connected Services tab

The application is deployed on Azure App Service and the database used is Cosmos DB, as presented in the following figure:

Figure 9.26 – Hosting and Service Dependencies

We can use the **Microsoft.Azure.Cosmos** packages and the **CosmosClient** class here. However, if you use Entity Framework as an object-relational mapper in the data layer, you can use the **Microsoft. EntityFrameworkCore.Cosmos** package. For more details, you can follow this link: `https://learn. microsoft.com/en-us/ef/core/providers/cosmos/?tabs=dotnet-core-cli`.

Summary

In this chapter, we learned how to create an Azure Cosmos DB account, and we created a database and container using **Data Explorer**. We explored the different API models, such as Core (SQL), which is recommended for new applications, the Azure Cosmos DB API for MongoDB, Cassandra, Azure

Table Storage, and Gremlin (Graph). We created resources by using the Microsoft .NET SDK v3. We connected the application to an Azure Cosmos DB account using the endpoint and the primary key. We used `CosmosClient` to create a database, create a container, and retrieve data from a document. We also connected Azure App Service with Azure Cosmos DB.

In the next chapter, we will explore big data storage. We will define Azure Data Lake Storage and learn how to create a more secure, high-performance framework for data analytics. If you need more information related to Cosmos DB, you can check out this link: `https://learn.microsoft.com/en-us/azure/cosmos-db/introduction`.

Questions

Before creating a container, which Azure Cosmos DB SQL API resource should you create first?

10

Big Data Storage Overview

Companies that use modern systems generate large volumes of heterogeneous data. This data must be exploited for marketing reasons or to make internal improvements to a product. This heterogeneous data demonstrates that a single data store is generally not the best approach.

It is also recommended to store different types of data in different data stores so that each one is geared toward a specific workload or usage pattern. If we use a combination of different data storage technologies, we are using what is called **polyglot persistence**. It is important to understand what Azure offers as a service for storing data warehouses and how we can use and analyze all that data.

In this chapter, we will explore big data storage and define Azure Data Lake Storage scalability, security, and cost optimization work. You will learn how to create a more secure, high-performance framework for data analytics.

In this chapter, we're going to cover the following main topics:

- Exploring Azure Data Lake Storage
- Exploring Azure Data Factory
- Exploring Azure Databricks
- Exploring Azure Synapse Analytics
- Exploring Azure Analysis Services

Exploring Azure Data Lake Storage

Azure Storage includes five Azure services:

- Azure Blob Storage, which is scalable storage for unstructured data
- Azure Queue Storage, which allows us to build a reliable queue of messages
- Azure Table Storage, which provides support for storing structured data
- Azure Files Storage, which is used for cloud-based file shares
- Azure Disks Storage, which provides managed storage for virtual machines disks

A data lake is a form of file storage, typically on a distributed filesystem for high-performance data access. The technologies that are commonly used to process queries against stored files and return data for reporting and analysis are Spark and Hadoop. These systems rely on a read-schema approach, which defines a tabular schema for semi-structured data files, where the data is parsed as it is read and no restrictions are applied when it is saved. Data lakes are ideal for supporting a mix of structured, semi-structured, and unstructured data that you want to analyze to avoid applying schemas when the data is written to storage.

Azure Data Lake Gen2 technology is created on top of Azure Blob Storage and supports most Blob Storage features. More benefits are related to Azure Data Lake, such as supporting the hierarchical namespaces, which means that we can store data in a file-like structure. This benefit improves the performance of the directory-managed operations to provide better support for a large-scale analytical engine.

In this section, we will start by exploring Azure Data Lake Storage. Azure Data Lake Storage combines the power of a high-performance filesystem with a large-scale storage platform to give you quick insights into your data.

Azure Data Lake Storage Gen2 builds on Azure Blob Storage features to optimize it, especially for analytical workloads. If you've used Hadoop Distributed File System, we can treat the data in the same way. With this feature, you can store your data in one place and access it through various Azure compute services, such as Azure Databricks, HDInsight, and SQL Data Warehouse. Data Lake Storage Gen2 supports ACLs and POSIX permissions. We can set a granular level of permissions at the directory or file level for data stored within the data lake. Azure Data Lake Storage organizes stored data by following a hierarchy of directories and subdirectories, such as a filesystem for easy navigation, which improves the performance of each directory-managed operation. This organization allows us to provide better support for a large-scale analytical engine, improved querying performance using data partitioning, and better file and data management.

The cost of storing data in Azure Data Lake Storage is the same as storing data in Azure Blob Storage. There is a slight increase in transaction costs for data access, but in many cases, these cost increases must be balanced against the need for fewer transactions due to more efficient queries. Overall, Azure Data Lake Storage allows you to build secure, enterprise-scale data lakes. It offers many of the benefits of Azure Blob Storage while doing so at minimal additional cost.

A good use case for Blob storage is archiving infrequently used data or storing website assets such as images and media. Azure Data Lake Storage Gen2 plays a fundamental role in any Azure data architecture. These architectures include building modern data warehouses, advanced analytics on big data, and real-time analytics solutions.

Common to all architectures, there are four phases of processing big data solutions:

1. The ingestion phase identifies the technologies and processes used to ingest the source data. This data can come from files, logs, and other types of unstructured data that need to be placed in Data Lake Storage.

2. The storage phase identifies where to store the recorded data. In this case, we use Azure Data Lake Storage Gen2. This phase is presented in the following figure:

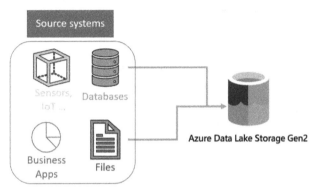

Figure 10.1 – Ingestion phase – Azure Data Lake Storage Gen2

3. The preparation and training phase identifies the technologies used to conduct, prepare, and model the training and subsequent scoring of the data science solution. Common technologies used in this phase are Azure Databricks, Azure HDInsight, and Azure Machine Learning.

4. Finally, the model and deployment phase includes the technology that presents data to users. This includes visualization tools such as Power BI and data stores such as Azure Synapse Analytics, Azure Cosmos DB, Azure SQL Database, and Azure Analysis Services. Combinations of these technologies are often used, depending on business needs.

Common file formats for general-purpose storage in Azure Data Lake include Excel, XML, JSON, binary, Avro, delimited, **Optimized Row Columnar (ORC)**, and Parquet.

We will create a new Azure Data Lake solution using the Azure portal and the Azure CLI in the following subsections.

Creating an Azure Data Lake instance using the Azure portal

Azure Data Lake Storage is built on top of Azure Storage; so, we will create a new storage account and enter the different information needed in the **Basics** tab.

In the **Advanced** tab, check **Enable hierarchical namespace** under **Data Lake Storage Gen2**. This setting will convert the standard Blob storage account into an Azure Data Lake storage account:

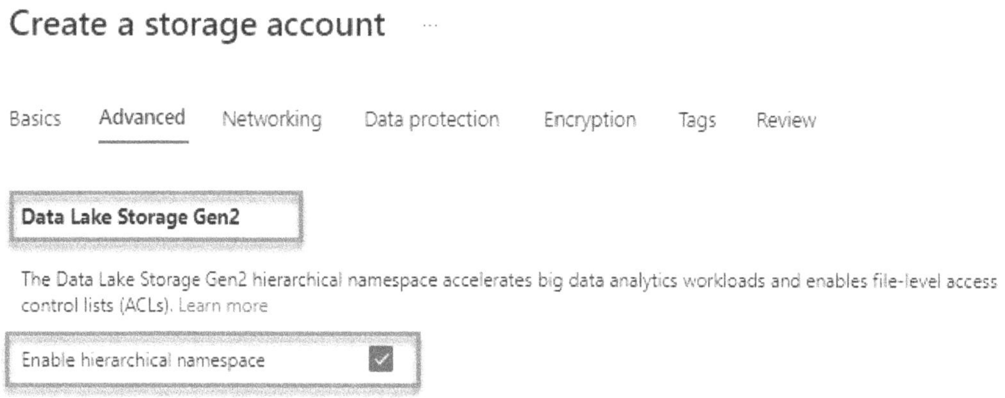

Figure 10.2 – Creating an Azure Data Lake instance using the Azure portal

Select **Review + Create** to review your storage account settings and create the account.

To enable Data Lake Storage capabilities on an existing account, open your storage account; then, under **Settings**, select **Data Lake Gen2 upgrade**. The **Upgrade to a storage account with Azure Data Lake Gen2 capabilities** page will be displayed, which includes three steps. The first step consists of reviewing the account changes before the launch of the upgrade and the second step will validate the different features that cannot be supported with Azure Data Lake Storage Gen2. We will get a list of discrepancies. The last step starts the upgrade once the validation succeeds:

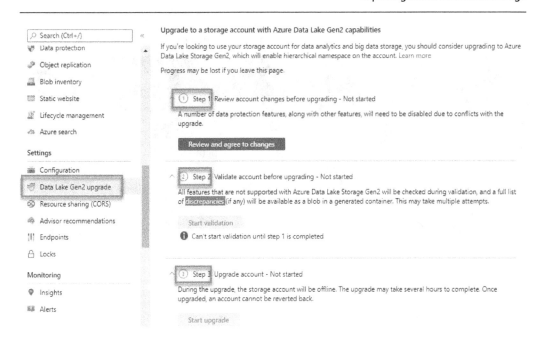

Figure 10.3 – Upgrading to a storage account with Azure Data Lake Gen2 capabilities

After a successful migration, we will get the following message: **This account has successfully been upgraded to an account with Azure Data Lake Gen2**.

In the next section, we will create an Azure Data Lake instance using the Azure CLI.

Creating an Azure Data Lake instance using the Azure CLI

We will run the Azure Cloud Shell browser tool to execute the following steps:

1. Create a new resource group to deploy our Azure Data Lake Storage into, and add the name and the location:

```
$ az group create --name healthcareRG --location eastus
```

In the following figure, we will execute the previous command line to create a new resource group:

```
hamida@Azure:~$ az group create --name healthcareRG --location eastus
{
  "id": "/subscriptions/                                    b/resourceGroups/healthcareRG",
  "location": "eastus",
  "managedBy": null,
  "name": "healthcareRG",
  "properties": {
    "provisioningState": "Succeeded"
  },
  "tags": null,
  "type": "Microsoft.Resources/resourceGroups"
}
```

Figure 10.4 – The result of adding a new resource group

2. Create a managed identity:

```
$ az identity create -g healthcareRG -n user1
```

The output is shown in the following screenshot:

```
hamida@Azure:~$ az identity create -g healthcareRG -n user1
{
  "clientId": "40268d7a-7038-4037-97d3-5865ad988122",
  "id": "/subscriptions/                                    )/
  "location": "eastus",
  "name": "user1",
  "principalId": "64bd7ba9-7afe-424c-b08b-fc36178d74a4",
  "resourceGroup": "healthcareRG",
  "tags": {},
  "tenantId": "1800b09f-2bb6-425c-b16c-3d7dde051282",
  "type": "Microsoft.ManagedIdentity/userAssignedIdentities"
}
```

Figure 10.5 – The result of creating a managed identity

3. Create a new storage area with Data Lake Storage Gen2, but before that, add an extension to the Azure CLI so that it can use the different features for Data Lake Storage Gen2:

```
$ az extension add --name storage-preview
$ az storage account create --name myhealthcarestorage1 \
    --resource-group healthcareRG \
    --location eastus --sku Standard_LRS \
    --kind StorageV2 --hns true
```

Once our deployment has finished, and we have the JSON output, we can go back to the portal and go to our resource groups. We will see that our new resource group has been created. Inside it, we have the Data Lake Storage account and the new user:

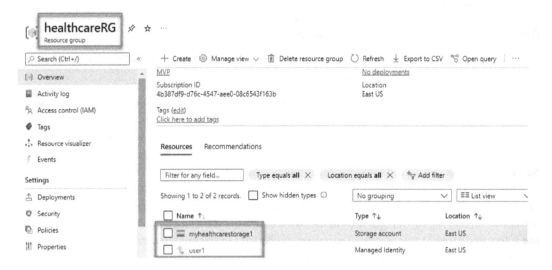

Figure 10.6 – Displaying a Data Lake Storage Gen2 resource

We can add the new user-assigned managed identity to the Storage Blob Data Contributor role on the storage account.

If we want to upgrade Azure Blob Storage with Azure Data Lake Storage Gen2 capabilities using the Azure CLI, we can run the following command:

```
$ az storage account hns-migration start --type validation -n
your-storage-account-name -g the-resource-group-name
```

In Azure, you can implement data ingestion at scale by creating pipelines that orchestrate the process of extracting, transforming, and loading data. This process is called ETL. A lot of work based on coordinating all of the sources, destinations, and transformations, as well as multiple databases, file formats, and data types, is often required, and this is the task of **Azure Data Factory** (**ADF**). You can use ADF to build and run pipelines. We will explore this service in the next section.

Exploring Azure Data Factory

ADF is an Azure cloud-based data integration service, a **Platform-as-a-Service** (**PaaS**) solution that allows you to orchestrate and automate data movement and data transformation. Data integration is the process of combining data from multiple sources and providing an integrated view of it.

ADF allows you to define and schedule data pipelines so that you can transfer and transform data. You can integrate your pipelines with other Azure services, making it easy for you to integrate data from cloud data stores. You can also process data using cloud-based compute resources and keep the results in another data store.

ADF is serverless, which means that you only pay for what you use.

With ADF, we can bring data into a common format. We can generate new insights we'd probably never have by keeping this data separately.

ADF is enterprise data ready because it has almost 90 different data connectors where we don't need to use a server. It is also a code-free transformation capability because we have mapping data flows, which provide a different transformation type that allows us to modify data. With ADF, we can run code on any Azure compute instance for any hands-on data transformations. It includes multiple **SQL Server Integration Services** (**SSIS**) packages that we can run on Azure, which optimize the ability to rehost on-premises databases in ADF. ADF uses Azure DevOps and GitHub to manage the data pipeline operations. We can use built-in multiple activities to simplify data publication to any data warehouse, database, or BI analytics engine. ADF simplifies data pipeline operations by allowing the automated provisioning of simple templates. It provides secure data integration because it uses a virtual network manager to prevent data exfiltration.

ADF components

We need to understand the key terms and basic concepts in ADF. There are six main components, which we will cover in this section, that we have to create our projects: pipelines, activities, datasets, linked services, data flows, and integration runtimes. Then, we will explore control flows, pipeline runs, triggers, parameters, and variables:

- **Pipelines** are logical groups of activities that perform a single unit of work. We can have more than one active pipeline. A group of activities in a pipeline performs a task. An activity can be scheduled in a sequence (run as a chain) or independently (run in parallel) within a pipeline. With pipelines, we can manage correlated activities in a single entity.

- **Activities** are processing steps inside a pipeline, as defined previously. Three types of activities are supported: data movement, so we can copy data from a store (source) to another store, which is the destination, data transformation, and activity orchestration.

- **Datasets** are data structures that give us a selected view into a data store, ideally to point to or reference the data used as inputs and outputs of a particular activity.

- **Linked services** are the connection strings that activities can use to connect to external services, which means that they are the connection information needed by ADF to connect to external resources, typically pointing to data sources (ingestion) or compute resources (transformation) required for execution.

- **Mapping data flows** create and manage data transformation logic graphs, which can be used to transform data of any size. We can build up a reusable library of data transformation routines to execute those scalable processes from our pipelines.

- **Integration runtime** is the compute infrastructure used by ADF. It provides fully managed data flows for the transformation and data movement processes, an activity dispatcher to route the service, and a manager that computes SSIS package execution tasks within the data pipelines.

The following are additional terminologies related to ADF:

- **Pipeline runs or pipeline execution**: We instantiate pipelines by passing arguments (values) to parameters (placeholders) that are defined by the activity pipeline. These arguments can be passed manually or using a trigger.

- **Trigger**: This is a unit of processing that determines when pipeline execution begins. We have different trigger types to use for different events.

- **Parameters**: These are key-value pairs of configuration that are used in read-only mode. They are populated from the runtime execution and defined in the pipeline. Dataset and linked services are strongly typed, referenceable, and reusable parameter entities.

- **Variables**: These are used within pipelines to store temporary values. States are used with parameters to pass values between activities, data flows, and pipelines.

- **Control flow**: This is the process of orchestrating the pipeline activities. It contains the chaining activities, defines parameters, branch activities, and custom state passing, and passes arguments for the pipeline running on demand or invoked by a trigger.

Now, we will create an ADF using the Azure portal.

Creating an ADF using the Azure portal

Open the Azure portal and select **Create a resource**. Then, follow these steps to create an ADF:

1. Select **Integration** and then click on **Create** under **Data Factory**.

2. On the **Basics** tab on the **Create Data Factory** page, the first part is related to the project details, so we will select the subscription and the resource group. Regarding the instance details, we will enter the name of our ADF instance, which must be unique, and the region where our ADF metadata will be stored. The version will be V2:

Create Data Factory   ...

Basics   Git configuration   Networking   Advanced   Tags   Review + create

Project details

Select the subscription to manage deployed resources and costs. Use resource groups like folders to organize and manage all your resources.

Subscription *   ⓘ   MVP   ⌄

Resource group *   ⓘ   healthcareRG   ⌄
Create new

Instance details

Name *   ⓘ   healthcareDF   ✓

Region *   ⓘ   East US   ⌄

Version *   ⓘ   V2   ⌄

Figure 10.7 – Create Data Factory – the Basics tab

3.   Configure a Git repository in Azure DevOps or GitHub. We can **Configure Git later**:

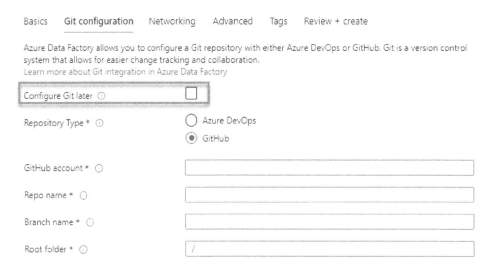

Create Data Factory   ...

Basics   Git configuration   Networking   Advanced   Tags   Review + create

Azure Data Factory allows you to configure a Git repository with either Azure DevOps or GitHub. Git is a version control system that allows for easier change tracking and collaboration.
Learn more about Git integration in Azure Data Factory

Configure Git later   ⓘ   ☐

Repository Type *   ⓘ   ◯ Azure DevOps
                            ⦿ GitHub

GitHub account *   ⓘ

Repo name *   ⓘ

Branch name *   ⓘ

Root folder *   ⓘ   /

Figure 10.8 – Create Data Factory – the Git configuration tab

4. Leave the default value as-is for the **Networking** tab. We can configure a private or public endpoint and choose whether or not to enable a managed virtual network. Under **Advanced**, we can enable encryption using a customer manager key to secure our ADF instance further.

5. Select **Review + create**, and then select **Create** after the validation has passed.

Once the ADF resource has been created, we must navigate to the ADF page and start exploring the different settings. Select **Open Azure Data Factory Studio**; a new page will be loaded in a separate tab in your browser:

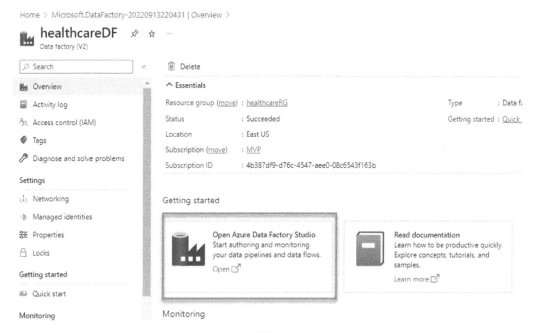

Figure 10.9 – Opening ADF Studio in the ADF service

To start with ADF, we will create a linked service, datasets, and pipeline.

Creating a linked service

We need to follow these steps:

1. To create a linked service to link a data store to our ADF instance, we can select any external store or Azure Storage, or a database, file, or NoSQL database. Select the **Manage** tab from the left pane and after that, select **+ New**:

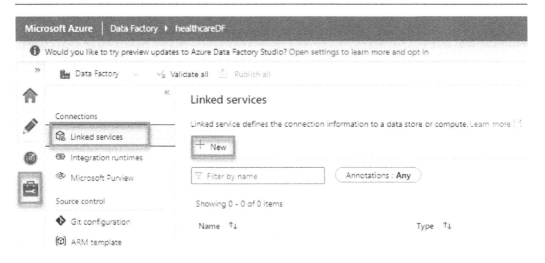

Figure 10.10 – Creating linked services

2.   Select **Azure Data Lake Storage Gen2** and then select **Continue**.

3.   Fill out the name, information related to the subscription, and the name of the storage account that we will use to store the data:

Figure 10.11 – Selecting the Azure Storage account to store data

4. Select **Test connect** to confirm that our ADF instance is connected to the storage account (Data Lake Storage Gen2). Then, select **Create** to save the linked service.

Creating a dataset

In this section, we will create two datasets: the first will be the input and the second will be the output. They will be of the Azure Data Lake Storage Gen2 type:

1. Select the **Author** icon on the left of the pane. Next, click on the plus icon (+) and select **Dataset**:

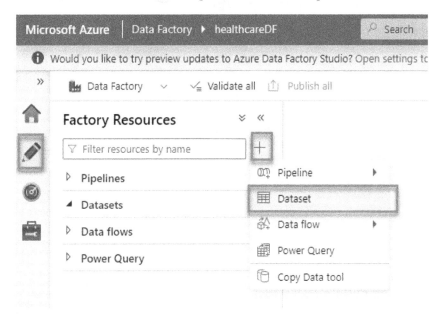

Figure 10.12 – Create a dataset

2. On the **New Dataset** page, select **Azure Data Lake Storage Gen2**. Then, select **Continue**.

3. Choose the format type of your data from **Avro**, **Binary**, **DelimitedText**, **Excel**, **JSON**, **ORC**, **Parquet**, or **XML**. To copy our files without parsing their content, we will select the **Binary** type.

4. Set the different properties on the **Set properties** page, namely, the name and the linked service. We will use the Data Lake Storage that we configured previously. Select a file path:

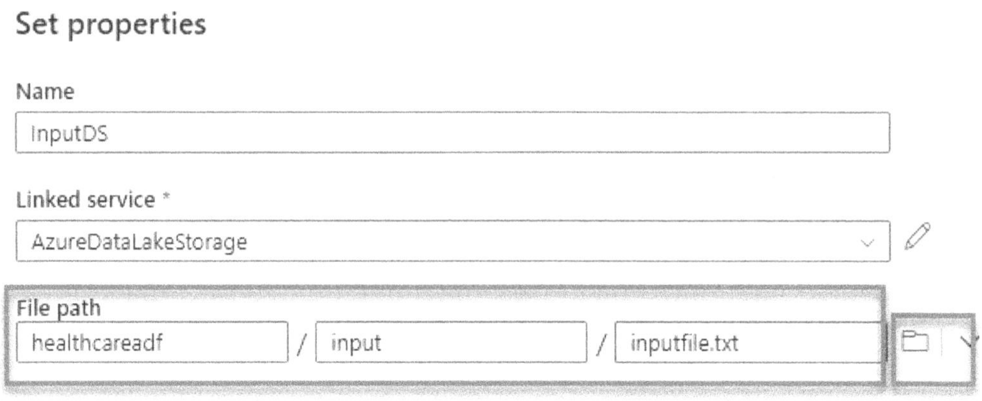

Set properties

Name

InputDS

Linked service *

AzureDataLakeStorage

File path

healthcareadf   /   input   /   inputfile.txt

Figure 10.13 – The Set properties page of a dataset for the input

5.   Create another dataset following the same steps as before, but specify this as an output. Select **OK** to confirm this:

Set properties

Name

OutputDS

Linked service *

AzureDataLakeStorage

File path

healthcareadf   /   output   /   File name

Figure 10.14 – The Set properties page of a dataset for the output

After creating the dataset, we will create our pipeline with a copy activity using the input and output datasets. This copy activity will copy our data from the input dataset file to the output dataset file settings.

Creating a pipeline

To create a pipeline, we will follow these steps:

1.   Again, select the plus (+) button, then **Pipeline | Pipeline**, as presented in the following figure:

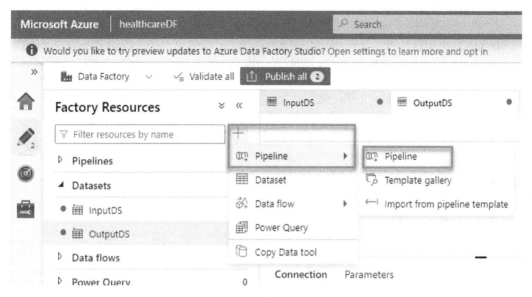

Figure 10.15 – Selecting a pipeline

Alternatively, we can select **Pipelines** and then **New pipeline**, as shown in the following figure, to create a new pipeline:

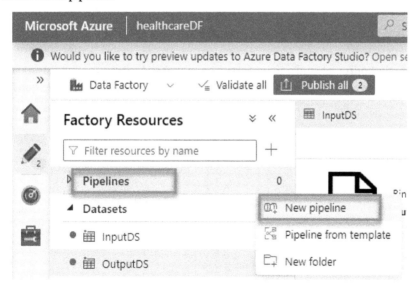

Figure 10.16 – Creating a pipeline

2. Name our pipeline `CopyPipeline`.

3. In the **Activities** toolbox, expand **Move & transform**. Next, drag the **Copy data** activity to the pipeline designer surface on the right. We will rename it `CopyFromDFToDF`, as follows:

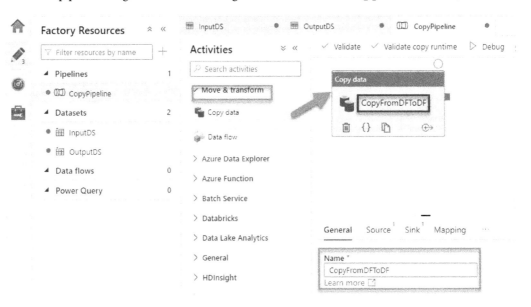

Figure 10.17 – Move & transform under Activities

4. Switch to the **Source** tab under the copy activity setting and select our input dataset (**InputDS**) for **Source dataset**. Then, switch to the **Sink** tab to link the output dataset (**OutputDS**).

In the following figure, you can see the input dataset configured in the activity settings:

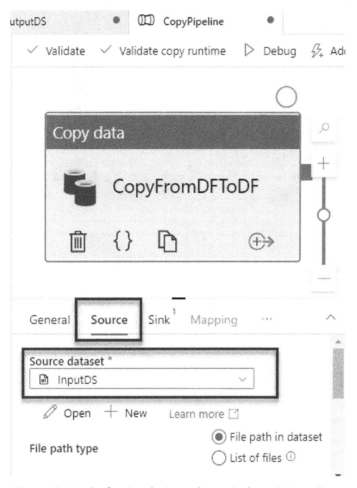

Figure 10.18 – Configuring the input dataset in the activity settings

In the following figure, you can see the output dataset configured in the activity settings:

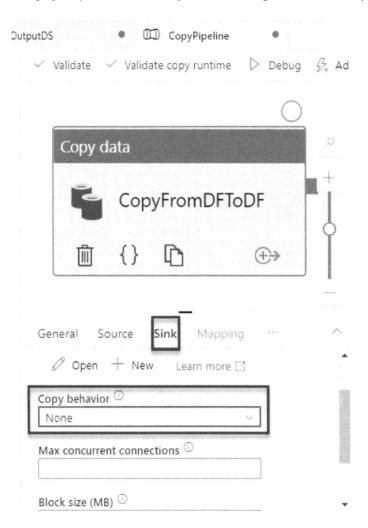

Figure 10.19 – Configuring the output dataset in the activity settings

5.   Select **Validate** on the pipeline toolbar to validate all the different settings we created previously. Once the pipeline has been validated without any errors, we will proceed with debugging our pipeline. So, select the **Debug** button. We will not publish any changes to the service in this phase. We will check the result in the **Output** tab of the pipeline settings.

6. We can trigger our pipeline manually. To do so, we need to publish the entities to ADF. So, select **publish** at the top. After that, select **Add Trigger** on the pipeline toolbar and then select **Trigger Now**.

7. We can automate our trigger on a schedule. In the **Author** tab, select **Add Trigger** on the pipeline toolbar. Next, select **New/Edit**. Select **Choose Trigger** on the **Add Triggers** page. After that, click on **New**. We will set up our trigger on the **New Trigger** page. After completing this configuration, we need to **Publish all** to apply the changes to ADF.

In this section, we used the Azure portal to create ADF, though we can also use the Azure CLI, Azure PowerShell, REST APIs, or a .NET application. You can read more about these options in the Microsoft documentation: `https://docs.microsoft.com/en-us/azure/data-factory/`.

A Data Factory is primarily a data orchestration and movement tool that calls other services such as Databricks to transform data. We will learn more about Azure Databricks in the next section.

Exploring Azure Databricks

In this section, we will explore the different features of Azure Databricks. However, we will start by defining Azure Synapse Analytics.

Azure Synapse Analytics

Databricks is a data, analytics, and AI company. This is a link to their platform: `https://www.databricks.com/`. They were the first creators of the open source versions of Apache Spark, Delta Lake, and MLflow. They used some of the architectural components of these services to compose Databricks. They tried to unify the innovation of the process within data science and data engineering.

Databricks offers an interactive workspace, helping to automate the production workflow. This entire workspace is fully managed, and Azure is the channel that Databricks uses to deliver a solution that can be deployed, regardless of the underlying infrastructure, such as servers and virtual machines.

Azure Databricks features

Azure Databricks offers multiples features:

- Exploiting the Spark engine, providing features such as machine learning, SQL, DataFrames, low-latency streaming APIs, as well as graph APIs to help find relationships in a dataset

- Several languages, such as Scala, Python, Java, R, and SQL, can be used in Azure Databricks

- Seamless and automatic integration of a workspace and Azure Active Directory

- Easy connection with ADF to create data pipelines

- Connecting the Databricks workspace to your Azure Storage so that you can use it as a filesystem where you can upload files to use in the workspace

- Adherence to industry security standards and compliance, such as FedRAMP and PCI DSS

We will explore the different components of Azure Databricks next.

Azure Databricks components

When we create a new Azure Databricks resource, we must create a new Databricks workspace. This is the first component; a workspace is an interactive tool used for exploring and visualizing data. In the workspace, we have Apache Spark clusters. They are used to serve as a compute engine to run the workloads in Azure Databricks. We can share clusters among users in a workspace. A cluster is auto-scalable, which means that we can allocate resources based on the component of the running job. We can auto-terminate any inactive cluster after a specific period of inactivity. We also have notebooks. These are part of the workspace, allowing us to read, write, query, visualize, or explore a dataset. We can connect a notebook to a cluster to run items. We have more options in notebooks, such as saving, sharing, importing, and exporting them in the workspace. Now, we will create a new Databricks workspace using the Azure portal.

Creating an Azure Databricks workspace

In this section, to set up our Azure Databricks environment, we will create an Azure Databricks cluster and notebooks, and walk through how to use them.

Open the Azure portal; we will start by clicking the **Create a resource** button. Then, in the search box, type in `Azure Databricks`. We will click on the **Create** button to create our resource. Follow these steps:

1. On the creation page, we must go through the various steps. We will start with the **Basics** tab. The first part is common to all resources: filling in the information for the subscription and the resource group. Regarding the instance details, we will introduce the workspace name and select the region and pricing tier. Under **Pricing Tier**, we have three options:

   - **Standard**, where we can use Apache Spark and integrate Azure Active Directory

   - **Premium**, where we have another feature that involves the use of role-based access controls

   - **Trial**, which allows us to use the Premium option for free for 14 days

2. On the **Networking** and **Advanced** tabs, we can use the default values. Select **Review + Create**. Once the validation has passed, click on the **Create** button.

3. Once the resource has been created, we will explore the Azure Databricks service page. Select **Launch Workspace**:

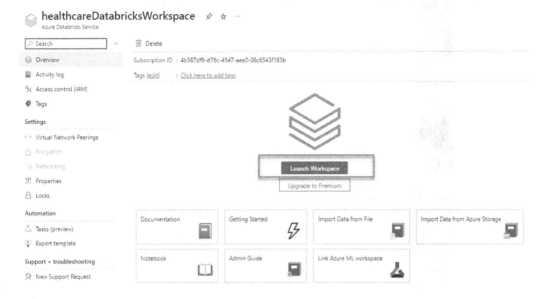

Figure 10.20 – Launching an Azure Databricks workspace

We will be redirected to another page in the browser.

4. On the landing page of the Databricks workspace, we will be asked to select the reason for using Databricks. We will select **Building data pipelines (ETL, streaming)**.

5. Create a new cluster that will act as the compute engine for any of the code we will write. Click on **Create a cluster**, and then select the **Create a cluster** button:

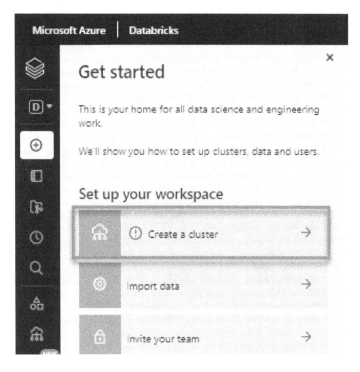

Figure 10.21 – Creating a cluster in a Databricks workspace

6. Let's fill out the different options with some simple configurations, as seen in the following screenshot, and select **Create Cluster** to confirm them:

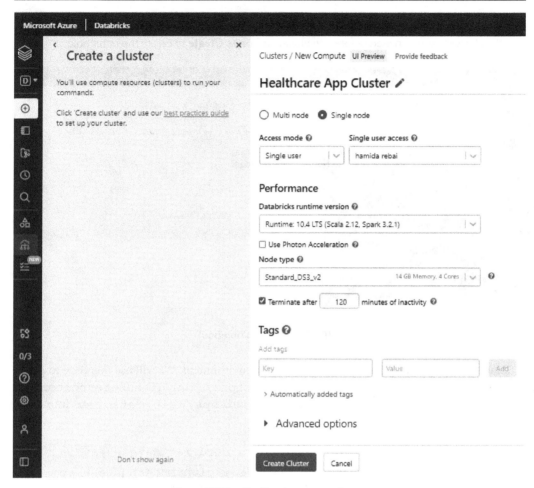

Figure 10.22 – Configuring a new cluster

7. Create a new notebook. We will select **Create** in the left pane, then **Notebook**, and then type a name.

8. The next option is to select the language you want to code in. We'll select the cluster, which is the cluster that we created before. Go ahead and press **Create** to create the notebook:

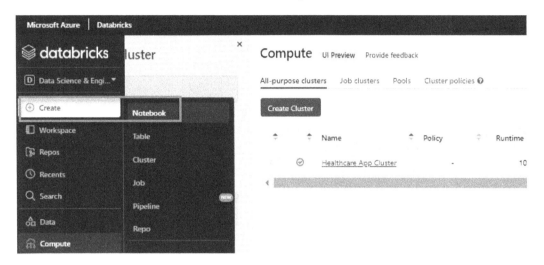

Figure 10.23 – Creating a notebook

We can use Python to bring data into the Databricks environment. We will use this code to pull data from the CSV file saved in a GitHub repository (https://raw.githubusercontent.com/didourebai/developerGuideForCloudApps/main/data.csv) into our Databricks environment:

```
import pandas as pd ## this import is used for data analysis
casesdatacsv = pd.read_csv('https://raw.githubusercontent.com/
didourebai/developerGuideForCloudApps/main/data.csv');
print ("First Databricks Table")
display(casesdatacsv)
```

The first line of this code is required. We call pandas, which is a Python package used for data analysis. The next line will grab a CSV file that contains some data from the GitHub repository. What it's going to do is save this as a DataFrame. We will call the DataFrame casesdatacsv. If we want to add a header to the result, we can use print with a message. Then, we can click on **Run Cell** or use *Shift + Enter*. The result is as follows:

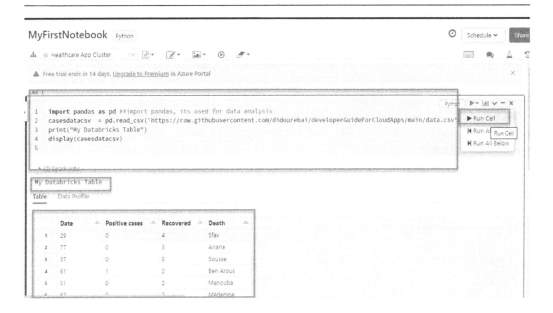

Figure 10.24 – Result of pushing data into a Databricks environment

We can expand the result if we have many rows in the displayed table.

Azure includes more solutions for data analytics. In the next section, we will explore Azure Synapse Analytics.

Exploring Azure Synapse Analytics

Azure Synapse Analytics is an integrated end-to-end solution for data analytics at scale. It includes multiple technologies and capabilities to combine the data integrity and reliability of a scalable, high-performance SQL Server-based relational data warehouse with the flexibility of a data lake and open source Apache Spark. It also contains native support for log and telemetry analytics through the use of Azure Synapse Data Explorer pools and built-in data pipelines for data ingestion and transformation.

All Azure Synapse Analytics services can be managed from a single interactive user interface called **Azure Synapse Studio**. It allows you to create interactive notebooks that can combine Spark code and Markdown content. Synapse Analytics is ideal for building a single, unified analytics solution on Azure.

Exploring Azure Analysis Services

Azure Analysis Services is a fully managed PaaS. It delivers cloud-oriented, enterprise-grade data models. In Azure Analysis Services, we can use advanced features of mashup and modeling to combine our data from more data sources into a single trusted tabular semantic data model. We can also define metrics and secure our data.

Summary

In this chapter, we covered various Azure data services that we can use. We used ADF and its data flows to record data. Ingested data can be stored in raw format in Azure Data Lake Storage Gen2. This raw data can be processed using various services, such as Azure Databricks, Azure HDInsight, and dedicated SQL pools. Managing individual services can be difficult. That's where Azure Synapse Analytics comes in. There are well-integrated services that work across all layers of a modern data warehouse, all from one environment.

In the next chapter, we will cover the **continuous integration/continuous deployment (CI/CD)** of containers on Azure. We will set up CD to produce our container images and orchestration.

Further reading

If you need more information related to big data storage in Azure, go to `https://learn.microsoft.com/en-us/azure/architecture/data-guide/technology-choices/data-storage` and `https://learn.microsoft.com/en-us/azure/storage/common/storage-introduction`.

Questions

Answer the following questions to test your knowledge of this chapter:

1. What is the difference between Azure Databricks and ADF?
2. How do we create an ADF instance using the Azure portal?

Part 3:
Ensuring Continuous Integration and Continuous Container Deployment on Azure

In this part of the book, we will focus on the continuous integration and continuous deployment of containerized applications to deploy to Azure containers and orchestrator cloud services.

This part comprises the following chapter:

- *Chapter 11, Containers and Continuous Deployment on Azure*

11

Containers and Continuous Deployment on Azure

Containers are a simple way for developers to build, test, deploy, update, and redeploy applications to a variety of environments, from a developer's local machine to an on-premises data center, and even to the cloud across multiple vendors.

Azure offers multiple services to deploy containers. Containers make it easy to continuously build and deploy your apps. With Kubernetes orchestration in Azure using Azure Kubernetes Service, our clusters of containers are replicable and easy to manage.

In this chapter, we will cover the **continuous integration and continuous delivery (CI/CD)** of containers on Azure. We will set up continuous deployment to produce our container images and orchestration.

In this chapter, we're going to cover the following main topics:

- Setting up continuous deployment for Docker with Azure DevOps and Azure Container Registry
- Continuous deployment for Windows containers with Azure DevOps
- Integrating Docker Hub with the CI/CD pipeline

Setting up continuous deployment for Docker with Azure DevOps and Azure Container Registry

In this section, we will set up a CI pipeline for building containerized applications. We will build and publish Docker images to Azure Container Registry using Azure DevOps to increase the reliability and speed of the deployment.

We will use our solution in GitHub. You can fork or clone it from this repository: `https://github.com/didourebai/healthcaresolution.git`. We will deploy our previous healthcare solution, which includes an ASP.NET API and two applications using ASP.NET MVC, and follow the steps illustrated in the following figure to deploy the application to Azure Container Registry. We will use Azure DevOps to ensure CI/CD:

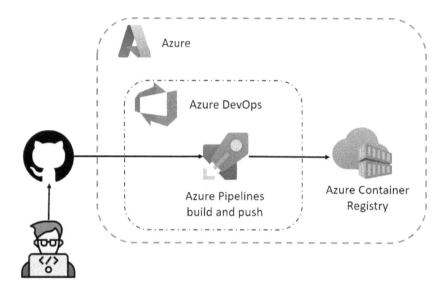

Figure 11.1 – Scenario: Build and push a Docker image to Azure
Container Registry using Azure DevOps pipelines

Note that we can use Azure Repos instead of a GitHub repository. A Container Registry resource is already created.

Creating the pipeline

In this section, we will create a pipeline as we need an Azure DevOps organization and a project already created. To do this, follow these steps:

1. Sign in to your Azure DevOps organization and navigate to your project: **healthcaresolution**.
2. Select **Pipelines** then **Pipelines**, select **Create Pipeline**, then select **New pipeline**, which will be displayed at the bottom, to create a new pipeline, as presented in the following figure:

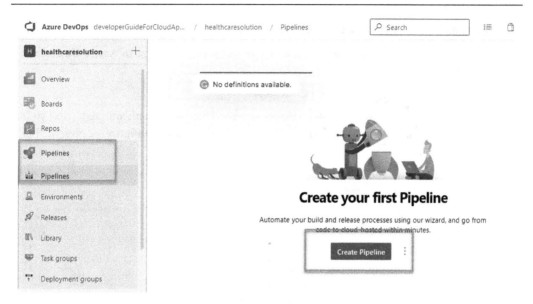

Figure 11.2 – Create pipeline

3. Connect to our remote repository that is in GitHub. We can select our code from any remote repository, for example, Azure Repos, Git, Bitbucket Cloud, or any other Git repository. In our case, we will select **GitHub (YAML)**. We need to select **Authorize Azure Pipelines** so that we have the necessary permissions to access our repository:

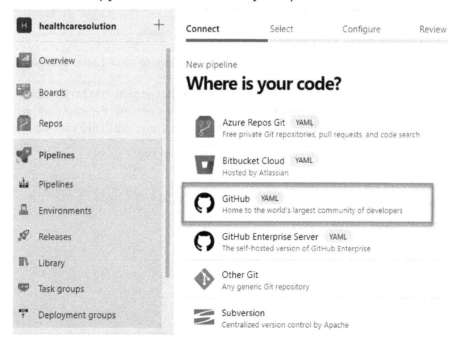

Figure 11.3 – Select the GitHub repository

4. After adding your GitHub credentials, select your repository from the list.

5. In the **Configure** tab, select the **Docker: Build and push an image to Azure Container Registry** task.

6. Select the subscription, then click on **Continue**.

7. You will be requested to connect to your Azure account. Once you're authenticated, select your container registry. Enter an image name, then select **Validate and configure**, as presented in the following figure:

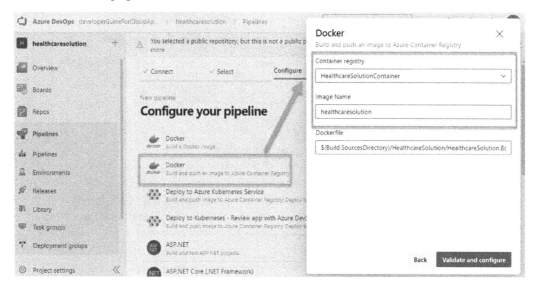

Figure 11.4 – Configuring the building and deploying of an image to Azure Container Registry

When Azure Pipelines builds your pipeline, it will create the connection to the Docker Registry service so your pipeline will be able to push images to Azure Container Registry. It will generate, in the end, a **YAML** file called `azure-pipelines.yml` that will define your pipeline.

8. Review your generated YAML pipeline and select **Save and run**:

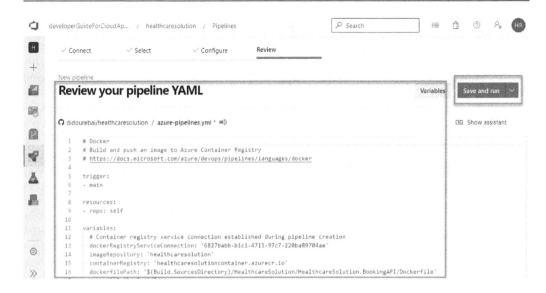

Figure 11.5 – The azure-pipelines.yml file

In our YAML file, we have the following:

- Variables that include the Container Registry connection, such as the image repository, the container registry, and the tag

- The VM image name, which is the OS environment

- The stages, which include the jobs for the build and the different settings

In the following figure, we can see the `azure-pipeline.yml` file and the different settings of **Stages**:

```
didourebai/healthcaresolution / azure-pipelines.yml *

30       steps:
         Settings
31       - task: Docker@2
32         displayName: Build and push an image to container registry
33         inputs:
34           command: buildAndPush
35           buildContext: $(Build.Repository.LocalPath)
36           repository: $(imageRepository)
37           dockerfile: $(dockerfilePath)
38           containerRegistry: $(dockerRegistryServiceConnection)
39           tags: |
40             $(tag)
```

Figure 11.6 – Settings of the stages in a YAML file

In the settings `.yml` file of the stages, we have the following:

- `Docker@2`: This task is used to build and deploy Docker containers.
- `command`: This indicates what we need to do, that is, the feature to run. In our case, we will use `buildAndPush`.
- `repository`: This indicates the name of our selected repository.
- `dockerfile`: This indicates the path of our Dockerfile.
- `containerRegistry`: This indicates the name of the container registry connection that we will use.
- `tags`: We will specify which tags need to apply to our container image.

9. Add a message in **Commit**, then select **Save and run**. In this step, we will commit all changes and the pipeline will be created.

The build job will start and will display **Queued** in the **Status** column. Click on the build and review your Azure DevOps pipeline logs. You can see the successful build and deployment of an image to Azure Container Registry:

Figure 11.7 – Build job in a running pipeline

Go to Azure Container Registry and in **Repositories**, under **Services**, you can check your image. After setting up continuous deployment for Docker with Azure DevOps and Azure Container Registry, we will see in the next section how we can set up continuous deployment for Windows containers using Azure DevOps.

Continuous deployment for Windows containers with Azure DevOps

In this section, we will deploy a containerized application to Azure App Service. The application is using .NET Framework. We will use a Windows container and push it to an Azure Container Registry. Next, we will deploy it to App Service.

In the following figure, we show how you can build and push an application using .NET Framework to Azure Container Registry and App Service:

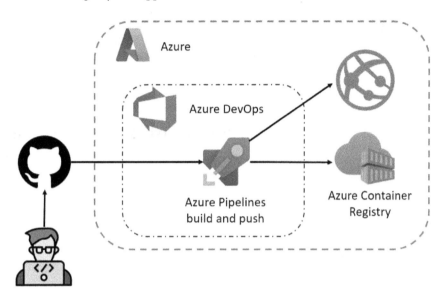

Figure 11.8 – Build and push an application to Azure Container
Registry and App Service using .NET Framework

Deploying to Azure Container Registry

We can follow the same steps presented previously to build and push an application to Azure Container Registry.

We can use another method to create the service connection between our project in Azure DevOps and Azure Container Registry using these steps:

1. Open your project in Azure DevOps and select **project settings** at the bottom of the page on the left. Under **Pipelines**, select **New Service connection** to define and secure the connection between the Azure subscription and your project in Azure DevOps using a service principal. Click on the **Create service connection** button:

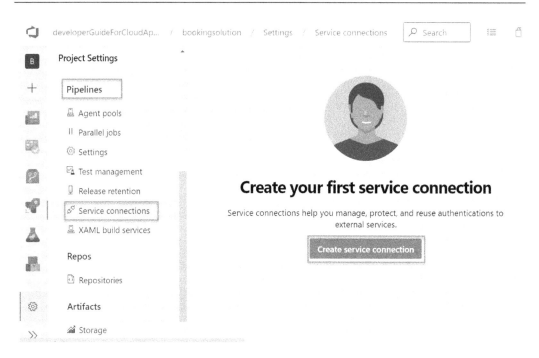

Figure 11.9 – Create a service connection between Azure DevOps and Azure services

2. A new dialog page will be displayed, requesting us to select a service connection from the list to link our Azure account to our Azure DevOps account. Select **Azure Resource Manager** and click on the **Next** button to continue the configuration:

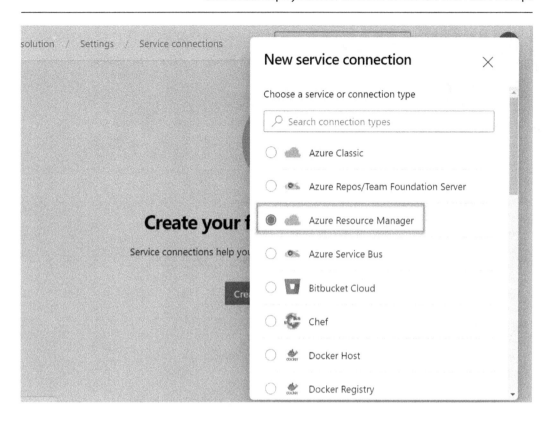

Figure 11.10 – Configure the new service connection

3. On the next dialog page, select **Service principal (automatic)** and click on **Next**. Select the subscription and the resource group, then enter a service connection. Select **Save** to finish the configuration.

Important note

If we only need to connect our Azure Container Registry with our DevOps account, we can select **New Service connection** to add a new connection link. Then, select **Docker Registry** and **Azure Container Registry**. Fill in all the parameters and select **Save** to confirm the connection.

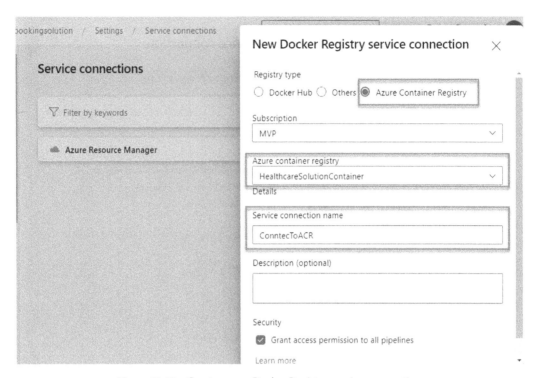

Figure 11.11 – Create a new Docker Registry service connection

4. Create a build pipeline by following the same steps as in the previous section. When the YAML file is generated, review it. Add variables that are used only after the creation of the pipeline. Click on **Variables**, then select **Add variable** and add all these variables:

- `imageRepository: your-image-name`

- `containerRegistry: 'your-registry-name.azurecr.io'`

- `applicationName: your-app-name`

5. Once the variables are saved, we will update vmImageName in the YAML file. Replace the default value, 'ubuntu-latest', with 'windows-latest'. We will also add buildContext after containerRegistry, as presented in the following screenshot, to make sure that all required application files are being copied over to the image filesystem. If we don't add this line, we will get an error indicating that the path inside of our Dockerfile build can't be found:

```
Settings
- task: Docker@2
  displayName: Build and push an image to container registry
  inputs:
    command: buildAndPush
    repository: $(imageRepository)
    dockerfile: $(dockerfilePath)
    containerRegistry: $(dockerRegistryServiceConnection)
    buildContext: '.'
    tags: |
      $(tag)
```

Figure 11.12 – Pipeline YAML for Container Registry

6. Select **Save and run** to start your pipeline build.

7. Once your image is pushed to the registry, push your container to App Service. Your App Service instance should be created.

In the next section, we will deploy the application to Azure App Service by creating a new release pipeline.

Deploying to Azure App Service

To deploy our application to Azure App Service, we will create a new release pipeline.

To create a new release pipeline, follow these steps:

1. From the dashboard, select **Pipelines** and then **Releases**.

2. Select **New Pipeline**, and then select **Empty job**, as presented in the following figure:

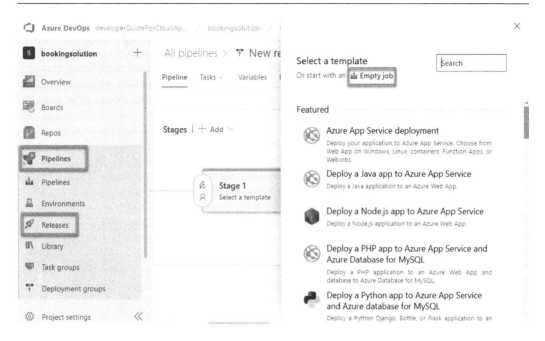

Figure 11.13 – Create a pipeline release in Azure DevOps

3. Add a new **Azure Web App for Containers** task to our stage and we will select **Azure App for Containers**:

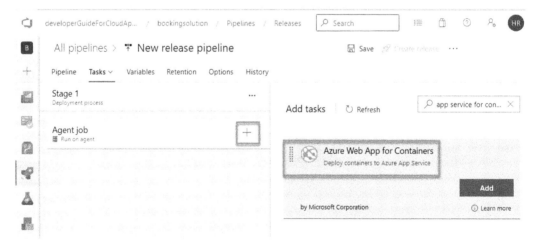

Figure 11.14 – Add a new task to the stage

4. Select the subscription and the application name and specify the fully qualified container image name. If you have a multi-container image, provide the names of them all in the **Image name** input. If you use a `docker-compose` file, add the path in the **Configuration File** input and select **Save**:

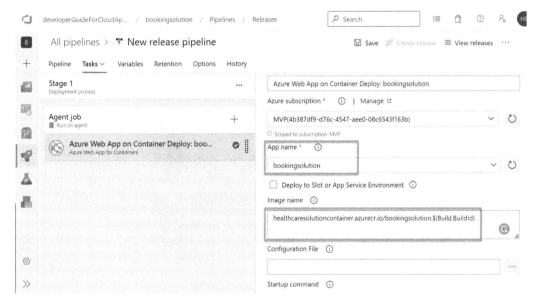

Figure 11.15 – Configure Azure Web App on Container Deploy

> **Important note**
>
> You will need to enable the admin user for your Container Registry. Select **Access keys** under **Settings** for your Container Registry and then select **Enabled** for **Admin user**. This option enables the use of the registry name as the username and the admin user access key as the password to allow Docker to log in to your Container Registry.

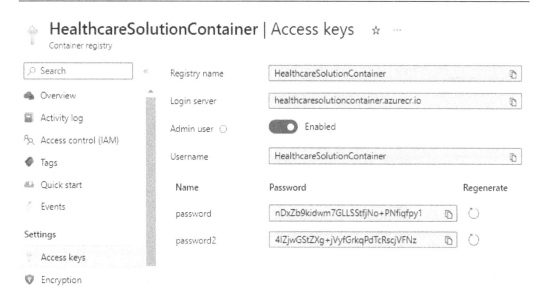

Figure 11.16 – Enable Admin user in Azure Container Registry

5.   On the **Pipeline** tab, select **Add an artifact**:

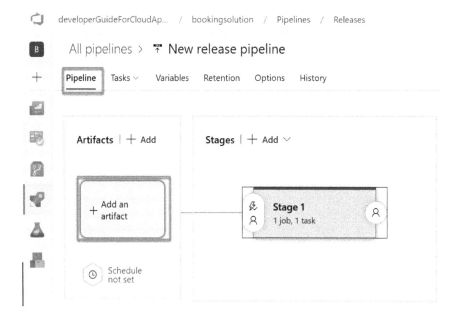

Figure 11.17 – Add an artifact

6. In **Add an artifact**, select **Build** for **Source type**, then select the project, the source, the default version, and select **Add**:

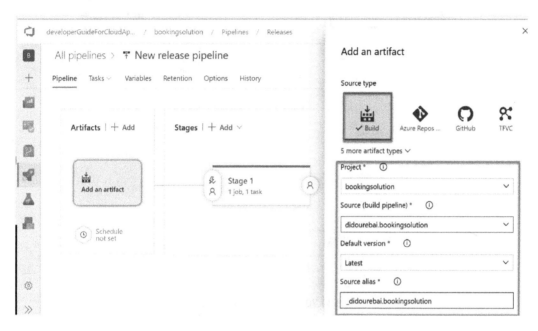

Figure 11.18 – Build source type

7. Enable **Continuous deployment trigger** and save all the modifications:

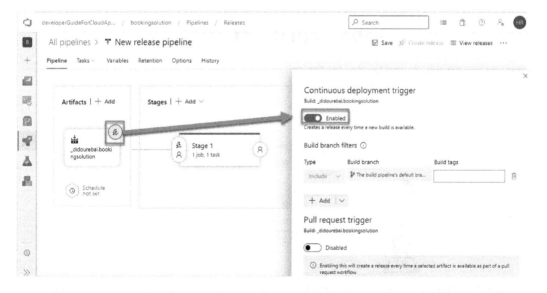

Figure 11.19 – Enable Continuous deployment trigger

8. Select **Create release**, add the stage and update the version for the artifact sources for this release. Then, click on the **Create** button:

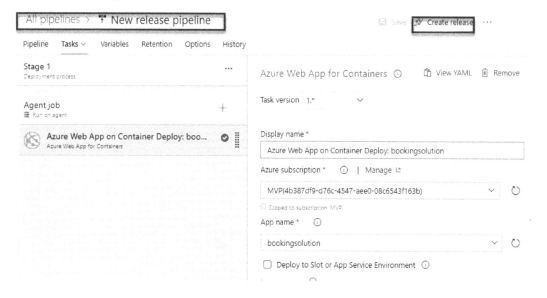

Figure 11.20 – Configure a release

We are ready to again push our local code to our repository to start the pipelines. We can add approval requests before updating App Service with the latest version.

We will see in the next section how we can integrate Docker Hub with the CI/CD pipeline.

Integrating Docker Hub with the CI/CD pipeline

In this section, we will set up a CI pipeline for building containerized applications. We will build and publish Docker images to Docker Hub:

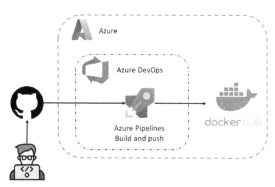

Figure 11.21 – Scenario: Build and push a Docker image to Docker Hub using Azure DevOps pipelines

We will start by adding a service connection to the Azure DevOps project and follow the same steps presented in the previous section.

In **Service connections**, add a new service connection to link Azure DevOps with the Docker Hub account. Select **Docker Registry**, and in **Registry type**, check **Docker Hub**. Complete the different settings, select **Verify** and **Save**:

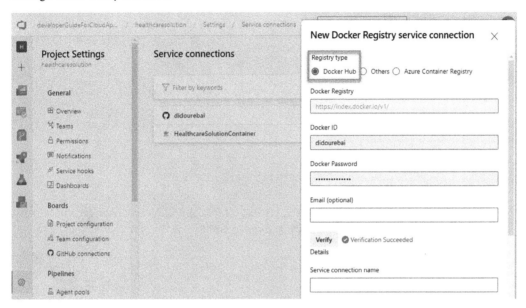

Figure 11.22 – Link Azure DevOps with Docker Hub Registry

We will follow the same steps as previously to create a build pipeline. Follow these steps:

1. To proceed with configuring your pipeline, select your repository and in the **Configure** tab, select **Docker- Build a Docker Image**. Then, select **Validate and configure**.

2. Add the following variables:

   - `imageRepository: your-image-name`

   - `containerRegistry: 'your-docker-hub-registry-name'`

   - `applicationName: your-app-name`

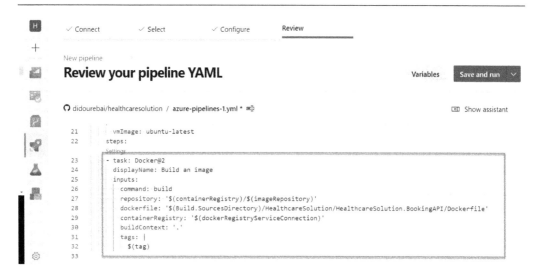

Figure 11.23 – Build pipeline to Docker Hub

3. Add a task to your pipeline; select **Docker**, then, select the connection service under **Container registry**, add the name of the repository, and select **Add** to save the task:

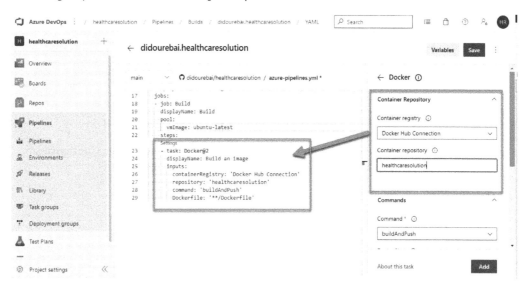

Figure 11.24 – Update a YAML file with the Docker Hub connection service

4. Select **Save and Run** after updating your YAML file to start your build pipeline. You can verify your repository in Docker Hub.

If you use an orchestrator, you can configure the deployment to Azure Kubernetes Service by editing the release pipeline:

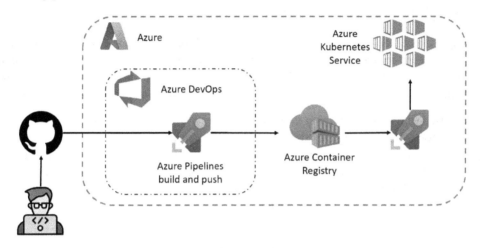

Figure 11.25 – Scenario: Build and push a Docker image to Azure
Kubernetes Service using Azure DevOps pipelines

We will add a new task to our stage. We can select **Deploy to Kubernetes** and configure the connection to the Azure Kubernetes Service resource, or we can use **Helm tool installer**, **Kubectl**, **Package and deploy Helm charts**. It depends on the application we're using:

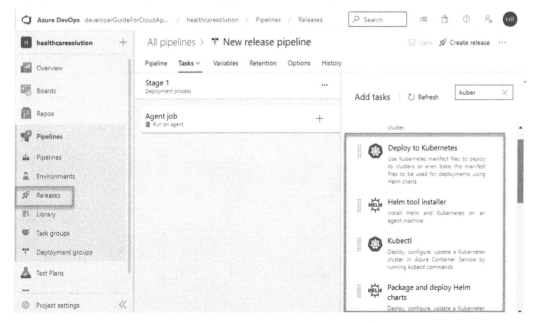

Figure 11.26 – Tasks in build pipelines to deploy Azure Kubernetes Service

We will save and run our release pipeline and our application will be pushed to Azure Kubernetes Service.

Summary

In this chapter, we learned about the build process and the deployment of a container using Azure DevOps. We pushed source code from GitHub to an Azure repository and configured the build pipeline to push our application to Azure Container Registry or Docker Hub for Linux and Windows environments. We created a release pipeline and published our application to **Web App for Containers**.

This book began with a basic definition of serverless and event-driven architecture and database as a service. We also worked through the different services in Azure, namely, Azure API Management using gateway pattern, event-driven architecture, Azure Event Grid, Azure Event Hubs, Azure message queues, function as a service using Azure Functions, and the database-oriented cloud. At every step along the way, you learned about creating, importing, and managing APIs and Service Fabric in Azure, and how to ensure CI/CD in Azure to fully automate the software delivery process (the build and release process).

Now, at the end of this book, you'll be able to build and deploy cloud-oriented applications using APIs, serverless architecture, Service Fabric, Azure Functions, and event-grid architecture.

Assessments

Chapter 1

1. What are the three key components of event-driven architecture?

   The key components of event-driven architecture are as follows:

   - **Event producers**: These generate a stream of events
   - **Event routers**: These manage event delivery between producers and consumers
   - **Event consumers**: These listen to the events

   For more details, see the *Understanding event-driven architecture* section.

2. What are the different Azure database services?

   The different Azure database services are as follows: Azure SQL Database, Azure SQL Managed Instance, SQL Server on Azure Virtual Machines, Azure Database for PostgreSQL, Azure Database for MySQL, Azure Database for MariaDB, Azure Cache for Redis, Azure Database Migration Service, and Azure Managed Instance for Apache Cassandra.

   For more details, see the *Exploring cloud databases* section.

Chapter 3

- Which packages need to be added to read messages from a Service Bus queue using a .NET Core application?

  `Azure.Messaging.ServiceBus`. For more details, see *Exercise 3 – publishing messages to a Service Bus queue using a .NET Core application*.

Chapter 4

1. Do we need to add more packages for Durable Functions?

   We need to add `Microsoft.Azure.WebJobs.Extensions.DurableTask`

2. What are *durable functions* in Azure?

   Durable Functions is an extension of Azure Functions that lets you write stateful functions in a serverless compute environment. For more details, see the *Developing durable functions* section.

Chapter 5

1. What is an Azure Service Fabric cluster?

    An Azure Service Fabric cluster is a network-connected set of **virtual machines** (**VMs**) in which your microservices are deployed and managed. For more details, see the *Clusters and nodes* section.

2. What is the difference between Kubernetes and Service Fabric?

    The main difference between Kubernetes and Service Fabric is that Service Fabric has three mechanisms to deploy and run applications: a propriety programming model that you can optionally use in your applications, the use of containers, and the use of any secure built with any platform and any language, with no modification required at all. For more details, see the *The differences between Service Fabric and Kubernetes* section.

3. How do we deploy a containerized application on Azure Service Fabric?

    To deploy a containerized application on Azure Service Fabric, we can use Visual Studio 2022 and select the Service Fabric template, and then we can select the container. For more details, you can see the *Building and executing a Docker container in Service Fabric* section.

Chapter 7

1. What is Azure SQL Database?

    Azure SQL Database is a hosted SQL database service in Azure. It runs on the SQL Server database engine. There are some important differences between Azure SQL Database and the traditional SQL Server. But most database administrators using SQL Server are able to migrate to Azure SQL Database. Azure SQL Database makes it extremely easy to scale a database. We are able to replicate a database in one or more locations around the world, which can improve performance if your application is used worldwide.

2. How do we connect Azure SQL Database to an ASP.NET app?

    To connect Azure SQL Database to an ASP.NET app, we start by creating and configuring the database connection, and we use Visual Studio 2022 and SQL Server Object Explorer. For more details about the different steps, see *Exercise 4 – connecting Azure SQL Database to an ASP.NET app*.

Chapter 8

1. What are the common uses of file storage?

    Azure Files can be used to completely replace traditional on-premises file servers or **Network-Attached Storage** (**NAS**) devices. Popular operating systems, such as Windows, macOS, and Linux, can directly mount Azure file shares wherever they are in any location. Azure Files makes it easy to lift and shift applications to the cloud that expect a file share to store file applications or user data. Azure file shares can also be replicated with Azure File Sync to Windows servers, either on-premises or on the cloud. To ensure performance and distributed caching of data where it is used, shared application settings are stored, for example, in configuration files, and diagnostic data (such as logs, metrics, and crash dumps) in a shared location. We store the tools and utilities needed to develop or administer Azure virtual machines or cloud services. That has covered all the common use cases of file storage for Azure Files.

2.   When do you use Azure files versus blobs?

You can use Azure files to lift and shift an application to the cloud, which already uses the native filesystem APIs to share data between it and other applications running in Azure. You want to store development and debugging tools that need to be accessed from many virtual machines. Azure blobs, on the other hand, can be used if you want your application to support streaming and random-access scenarios and you want to be able to access application data from anywhere. There are a few other distinguishing features on when to select Azure files over Azure blobs. Azure files are true directory objects, while Azure blobs are a flat namespace. Azure files can be accessed through file shares. However, Azure blobs are accessed through a container. Azure files provide shared access across multiple virtual machines, while Azure disks are exclusive to a single virtual machine. For more details, see the *When to use Azure files versus blobs* section.

Chapter 9

1.   Before creating a container, which Azure Cosmos DB SQL API resource should you create first?

We will start by adding a new database and a container. After, we will add data to the created database. We will query the data and, in the end, use the SDK of Cosmos DB to connect to the Azure Cosmos DB SQL API while using C# as the programming language. Then, we will add a new database and a new container.

Chapter 10

1.   What is the difference between Azure Databricks and ADF?

Azure Data Factory (**ADF**) is an orchestration tool for data integration services to carry out ETL workflows and orchestrate data transmission at scale. Azure Databricks provides a single collaboration platform for data scientists and engineers to execute ETL and create machine learning models with visualization dashboards.

2.   How do we create an ADF instance using the Azure portal?

To create an ADF instance in the Azure portal, we will select Data Factory from the resources in the Azure portal, and then, we will follow the different steps mentioned in the *Creating an ADF using the Azure portal* section.

Index

`Packt.com`

Subscribe to our online digital library for full access to over 7,000 books and videos, as well as industry leading tools to help you plan your personal development and advance your career. For more information, please visit our website.

Why subscribe?

- Spend less time learning and more time coding with practical eBooks and Videos from over 4,000 industry professionals

- Improve your learning with Skill Plans built especially for you

- Get a free eBook or video every month

- Fully searchable for easy access to vital information

- Copy and paste, print, and bookmark content

Did you know that Packt offers eBook versions of every book published, with PDF and ePub files available? You can upgrade to the eBook version at `packt.com` and as a print book customer, you are entitled to a discount on the eBook copy. Get in touch with us at `customercare@packtpub.com` for more details.

At `www.packt.com`, you can also read a collection of free technical articles, sign up for a range of free newsletters, and receive exclusive discounts and offers on Packt books and eBooks.

Other Books You May Enjoy

If you enjoyed this book, you may be interested in these other books by Packt:

Azure for Developers - Second Edition

Kamil Mrzygłód

ISBN: 978-1-80324-009-1

- Identify the Azure services that can help you get the results you need
- Implement PaaS components – Azure App Service, Azure SQL, Traffic Manager, CDN, Notification Hubs, and Azure Cognitive Search
- Work with serverless components
- Integrate applications with storage
- Put together messaging components (Event Hubs, Service Bus, and Azure Queue Storage)
- Use Application Insights to create complete monitoring solutions
- Secure solutions using Azure RBAC and manage identities
- Develop fast and scalable cloud applications

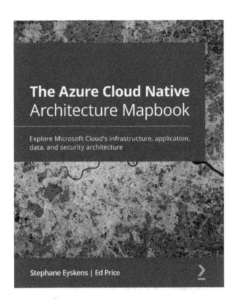

The Azure Cloud Native Architecture Mapbook

Stéphane Eyskens, Ed Price

ISBN: 978-1-80056-232-5

- Gain overarching architectural knowledge of the Microsoft Azure cloud platform

- Explore the possibilities of building a full Azure solution by considering different architectural perspectives

- Implement best practices for architecting and deploying Azure infrastructure

- Review different patterns for building a distributed application with ecosystem frameworks and solutions

- Get to grips with cloud-native concepts using containerized workloads

- Work with AKS (Azure Kubernetes Service) and use it with service mesh technologies to design a microservices hosting platform

Packt is searching for authors like you

If you're interested in becoming an author for Packt, please visit `authors.packtpub.com` and apply today. We have worked with thousands of developers and tech professionals, just like you, to help them share their insight with the global tech community. You can make a general application, apply for a specific hot topic that we are recruiting an author for, or submit your own idea.

Share your thoughts

Now you've finished *A Developer's Guide to Building Resilient Cloud Applications with Azure*, we'd love to hear your thoughts! Scan the QR code below to go straight to the Amazon review page for this book and share your feedback or leave a review on the site that you purchased it from.

`https://packt.link/r/1804611719`

Your review is important to us and the tech community and will help us make sure we're delivering excellent quality content.

Download a free PDF copy of this book

Thanks for purchasing this book!

Do you like to read on the go but are unable to carry your print books everywhere? Is your eBook purchase not compatible with the device of your choice?

Don't worry, now with every Packt book you get a DRM-free PDF version of that book at no cost.

Read anywhere, any place, on any device. Search, copy, and paste code from your favorite technical books directly into your application.

The perks don't stop there, you can get exclusive access to discounts, newsletters, and great free content in your inbox daily

Follow these simple steps to get the benefits:

1. Scan the QR code or visit the link below

https://packt.link/free-ebook/9781804611715

2. Submit your proof of purchase
3. That's it! We'll send your free PDF and other benefits to your email directly

www.ingramcontent.com/pod-product-compliance
Lightning Source LLC
Chambersburg PA
CBHW060519060326
40690CB00017B/3320